Reproductive Geographies

The sites, spaces and subjects of reproduction are distinctly geographical. Reproductive geographies span different scales—body, home, local, national, global—and movements across space.

This book expands our understanding of the socio-cultural and spatial aspects of fertility, pregnancy and birth. The chapters directly address global perspectives, the future of reproductive politics and state-focused approaches to the politicisation of fertility, pregnancy and birth. The book provides up-to-date explorations on the changing landscapes of reproduction, including the expansion of reproductive technologies, such as surrogacy and intrauterine insemination. Contributions in this book focus on phenomenologically inspired accounts of women's lived experience of pregnancy and birth, the biopolitics of birth and citizenship, the material histories of reproductive tissues as "scientific objects" and engagements with public health and development policy.

This is an essential resource for upper-level undergraduates and graduates studying topics such as Sociology, Geographies of Gender, Women's Studies and Anthropology of Health and Medicine.

Marcia R. England is Associate Professor of the Department of Geography at Miami University. Her research interests are in two main areas: access to public spaces and media/pop culture geographies.

Maria Fannin is Reader in Human Geography, in the School of Geographical Sciences at the University of Bristol. Her research interests include the social and economic dimensions of health, medicine and technology, particularly in relation to reproduction and women's health.

Helen Hazen is Teaching Associate Professor, Department of Geography and the Environment at the University of Denver, Colorado. Her research focuses on issues related to health and the environment.

Routledge International Studies of Women and Place

Series editors:
Janet Henshall Momsen, *University of California, USA* and **Janice Monk**, *University of Arizona, USA*

Feminism / Postmodernism / Development
Edited by Marianne H. Marchand and Jane L. Parpart

Women of the European Union
The Politics of Work and Daily Life
Edited by Maria Dolors Garcia Ramon and Janice Monk

Who Will Mind the Baby?
Geographies of Childcare and Working Mothers
Edited by Kim England

Feminist Political Ecology
Global Issues and Local Experience
Edited by Dianne Rocheleau, Barbara Thomas-Slayter, and Esther Wangari

Women Divided
Gender, Religion and Politics in Northern Ireland
Rosemary Sales

Women's Lifeworlds
Women's Narratives on Shaping their Realities
Edited by Edith Sizoo

Gender, Planning and Human Rights
Edited by Tovi Fenster

Gender, Ethnicity and Place
Women and Identity in Guyana
Linda Peake and D. Alissa Trotz

Reproductive Geographies
Bodies, Places and Politics
Edited by Marcia R. England, Maria Fannin and Helen Hazen

For more information about this series, please visit: www.routledge.com/
Routledge-International-Studies-of-Women-and-Place/book-series/SE0406

Reproductive Geographies
Bodies, Places and Politics

Edited by
**Marcia R. England, Maria Fannin and
Helen Hazen**

Routledge
Taylor & Francis Group

LONDON AND NEW YORK

First published 2019
by Routledge
2 Park Square, Milton Park, Abingdon, Oxon OX14 4RN

and by Routledge
52 Vanderbilt Avenue, New York, NY 10017

First issued in paperback 2020

Routledge is an imprint of the Taylor & Francis Group, an informa business

British Library Cataloguing-in-Publication Data
A catalogue record for this book is available from the British Library

Library of Congress Cataloging-in-Publication Data
A catalog record has been requested for this book

ISBN 13: 978-0-367-58265-4 (pbk)
ISBN 13: 978-0-8153-8619-3 (hbk)

Typeset in Times New Roman
by Taylor & Francis Books

Contents

Illustrations

Figures

Tables

Contributors

Marcia R. England is Associate Professor in the Department of Geography at Miami University. As an urban, cultural and feminist geographer, Marcia England's research interests are in two main areas: access to public spaces and media/pop culture geographies. She employs a feminist lens to examine social norms in real and reel spaces. This work includes a book on mediated representations of public and private spaces entitled *Public Privates: Feminist Geographies of Mediated Spaces*. She has also published on liminal spaces and abject bodies. Her teaching focuses on qualitative methods as well as geography and gender.

Maria Fannin is Reader in Human Geography at the University of Bristol, U.K. Her research interests include the social and economic dimensions of health, medicine and technology, particularly in relation to reproduction and women's health. Her current research focuses on the value attached to human placental tissue in the biosciences, medicine and alternative health practices. She has also carried out research on cord blood donation, placenta biobanking and philosophical approaches to "placental ethics". Her work has appeared in the academic journals *Body & Society, Feminist Theory* and *New Genetics & Society*.

Helen Hazen is Teaching Associate Professor at the University of Denver, Colorado. Her research focuses on issues related to health and environment. Most recently, she has begun to explore geographies of birth and specifically the influence of place on birth experiences. She has written a textbook on health geography and teaches on a wide variety of health-related topics.

Amita Bhakta is a Ph.D. student at the Water Engineering and Development Centre (WEDC) in the School of Architecture, Building and Civil Engineering at Loughborough University. She has particular interests in water, sanitation and hygiene (WASH) and gendered issues, with a specific focus on the equitable inclusion of perimenopausal women in relation to meeting their WASH-related infrastructural and social needs.

Dalia Bhattacharjee is a Ph.D. candidate in the Department of Humanities and Social Sciences at the Indian Institute of Science Education and

Research (IISER) Mohali, India. Her dissertation project is an ethnographic inquiry into the lives of the women working as surrogate mothers in India. The research, rooted in feminist geography and emotional geography, argues for the re-positioning of reproductive laborers in the commercial surrogacy industry, not as docile bodies, but as active agents.

Juliane Collard is a Ph.D. candidate in the Geography Department at the University of British Columbia. Her current research explores how life is taken hold of, valued and governed in assisted reproduction and reproductive medicine.

Carl T. Dahlman is Professor of International Studies at Miami University. His current research focuses on temporality and uncertainty, decentralisation and autonomy arrangements in multiethnic societies and transnational legal processes related to war crimes prosecution.

Jenna Dixon is a Postdoctoral Fellow in the Department of Geography and Environmental Management at the University of Waterloo. Her research interests centre broadly around four streams: health inequalities, health and development, gender and health and knowledge translation.

Julie Fisher is a lecturer at the Water, Engineering and Development Centre (WEDC), Loughborough University, concerned with issues of equity and inclusion related to water, sanitation and hygiene.

Robert Kaiser is a professor in the Geography Department at the University of Wisconsin–Madison. His research interests include political geographies of nationalism, borders and bordering and identification and differentiation. Recent work explores the performativity of place, identity and scale; political events, event space and anti-event space; citizenship and statelessness; the birth of cyberwar; and the current work on birth and biopolitics.

Katie Merkle is a feminist geographer and doctoral student in the Department of Geography at the University of Wisconsin-Milwaukee, U.S.A.

Brian Reed is a lecturer at the Water, Engineering and Development Centre, School of Architecture, Building and Civil Engineering, Loughborough University, with interests in how the technical aspects of water and sanitation engineering interact with socio-economic issues.

Andrea Rishworth is a Ph.D. candidate in the Department of Geography and Environmental Management at the University of Waterloo, Canada. Her research interests lie in health geography and the geography of aging, with a specific focus on gendered inequalities in health and wellbeing and environmental health linkages. Her current work addresses issues of health and wellbeing among aging populations in Uganda. She has also conducted research in Ghana on maternal health and health policy.

Risa Whitson is an Associate Professor in the Department of Geography and Women's, Gender, and Sexuality Studies Program at Ohio University. Her research interests include informal and nonstandard work, geographies of birth and reproduction, gender and development and feminist methodologies.

Acknowledgements

We would like to thank our series editors, Janet Momsen and Janice Monk, for inclusion in the Routledge International Studies of Women and Place series. We would also like to thank all the participants in our Reproductive Geographies sessions at the San Francisco and Boston meetings of the American Association of Geographers for starting the conversations that inspired this collection.

Introduction

A call for reproductive geographies

Maria Fannin, Helen Hazen and Marcia R. England

Daily media reports inform us of new forms of contraception, technological leaps in therapies for infertility, the loosening or tightening of restrictions on access to abortion, changes to state-mandated policies regarding family size, the ethics and economics of surrogacy, the persistence of racial disparities in maternal and child health, advances in embryonic stem cell research, informal exchanges of breast milk, the relationship between birthplace and rights of citizenship and yet other ways in which fertility, pregnancy and birth feature in people's lives. Many of these phenomena have pivotal connections to space and place—"reproductive geographies" are everywhere.

This edited collection brings together research and debates on the many geographies of reproduction, arguing that processes of fertility, pregnancy and birth benefit greatly from geographical examination. At a fine scale, the body itself is a critical site of geographic analysis, its "fleshy materiality" a rich site for exploration of affective, material and experiential dimensions of reproduction. Embodied experiences of natality, of giving birth, of being born, and of experiencing one's body as generative or procreative are lived differently in different places and times. Yet an exploration of the diversity of these experiences, across time and space, is seldom brought together in the geographical literature. In addition, social perspectives on bodies are central to understanding the enormous legal and policy changes over the course of the twentieth century that have made reproduction—viewed as an aspect of both collectivities or populations and as the property of the individual—one of the most salient frames through which to understand modern forms of power. A complex relationship exists between bodies and society, in which bodies are pathologised and yet nurtured, tied to both nature and science (Douglas 1966; Rose 1993; McDowell 1999).

A second important scale of analysis is the place of reproductive activities. Reproduction and new reproductive technologies have critical connections to particular places, especially through the notion of a "sense of place". Geographical exploration of reproductive experiences highlights how women's birth experiences are commonly tied to emotional and embodied senses of place, for instance, with different expectations and experiences tied to home-births versus hospital births (Abel and Kearns 1991; Fannin 2003; Kornelsen et al. 2010). More intimate geographies of reproduction suggest that places of insemination or conception are equally meaningful and place-specific. Particular places of the body itself are also filled with psycho-social significance, leading to expectations of appropriate behaviours. Taboos against exposing breasts, for instance, are clearly related to their sexualised nature, and yet have wide-reaching implications if we consider debates in many places related to where breastfeeding is deemed appropriate or inappropriate (Boyer 2012, 2018; Lane 2013).

At even larger scales, reproduction is intimately tied to the imagined futures of communities, nation-states and even the globe itself (Edelman 2004; Nahman 2013; Bashford 2014). That reproduction, whether collective or individual, is the target of state intervention is a modern phenomenon. The entry of the "life" of a collectivity or a subject into political calculation demonstrates how modern politics is also *bio*politics. When reproduction is envisioned as a problem of population, reproduction becomes inseparable from concerns over national security, economic productivity, resource use and geopolitical power. Attention to the ways social structures, cultural norms and state policies construct frameworks around what is and is not considered appropriate, desirable or even legal in the realm of reproduction is one of the most salient issues of our time. Reproduction, with its close connections to religious and cultural practices and landscapes, provides a complex case study of the ways in which societies attempt to manage individual rights and social goals, introducing complex geographies of power.

While we recognise that the term "reproduction" is itself an abstraction drawn from economic and scientific contexts that have historically de-emphasised concern for the rich and varied experiences of pregnancy and birth, we think it can be useful for drawing together the diverse empirical and conceptual approaches to these phenomena. We use it here to signal the importance of centring reproduction—as lived experience, as object of politics and policy—in geographical analysis. In this way, this book combines feminist and health geography approaches to pregnancy and birth with ecological and biopolitical concerns. In the following section, we situate the contribution of this volume in a broader discussion of current ideas about reproduction that have emerged from the social sciences.

Reproductive health and social justice

Reproductive health is an issue of social justice, encompassing not just the effort to exercise sovereignty over one's body but also the broader economic and institutional constraints that limit people's reproductive "choices". Abortion provides a particular flash point. In Brazil, for instance, where an estimated one in five women have had an abortion, there have traditionally only been three legal conditions for the procedure: rape, when the woman's life is at risk or if the foetus has a congenital brain disorder (Griffin 2018). Nonetheless, in November 2017, a Brazilian congressional committee voted in favour of a constitutional amendment that bans abortion in *all* cases. The bill is currently suspended due to other political crises but is among a variety of proposed laws that limit reproductive health options and autonomy. In the United States, the most stringent ban on abortion became law in Iowa in 2018. Called the "heartbeat bill", it bans abortion after a foetal heartbeat is detected (at about six weeks and before many women know they're pregnant). Exceptions are made only in cases of rape, incest or medical emergency. In other places, laws are increasing access to reproductive medicine, as illustrated in Northern Ireland by Belfast City Council's April 2018 vote to decriminalise abortion, and the Republic of Ireland's 2018 referendum on abortion that will likely lead to its legalisation in Ireland.

However, reproductive justice activists emphasise that struggles for individual choice related to reproduction obscure the wider context of people's differential access to the freedoms associated with reproductive rights. The exercise of (neo)liberal rights of individual freedom to choose is limited when access to healthcare is stratified by race, class and citizenship status (Roberts 1997; Silliman et al. 2004). New social justice issues are also arising as transgender people assert their reproductive rights. If testosterone is not used or is stopped, many transgender men have the capacity to become pregnant, generating complex social justice issues as pregnant people navigate transphobic healthcare institutions. Research suggests that societal censure may have practical implications for trans men who become pregnant, as they may have negative experiences due to a lack of medical assistance, particularly supportive prenatal care (Light et al. 2014).

More broadly, men seeking to control their own fertility have often been neglected in discussions of reproductive health. Traditionally, men have had access to few temporary male contraceptives other than the condom, although in March 2018 researchers stated they were on track with the development of a contraceptive pill for men (The Endocrine Society 2018). The new drug (dimethandrolone undecanoate or DMAU) showed marked suppression of testosterone levels as well as two hormones required for sperm production after daily ingestion. Technological developments in the provision of reproductive healthcare such as these are constantly changing the social and cultural dynamics of reproduction.

Technological reproduction and feminist politics

New reproductive technologies have been developing at a rapid rate since the introduction of the contraceptive pill and other forms of hormonal birth control in the mid to late twentieth century, followed by a similar expansion in fertility interventions in the latter part of the twentieth century. These technologies have led to dramatic changes in the reproductive landscape, which have become closely intertwined with feminist politics as access to new technologies is debated. Many have argued that the medicalisation of women's bodies has left women largely "absent" from discussions of reproductive technologies, with women instead often reduced to mere body parts— uterine environments (Berg 1995) or the "host" for a foetus (Somashekhar and Wang 2017). Vocal abortion debates persist in many communities, tied to differences in perceptions of the relative value of maternal and foetal rights. Contraceptive debates continue in many places, related to the degree to which women have the right to determine their reproductive futures.

Similarly, access to fertility technologies is viewed as central to the extension of women's reproductive rights, although their development in the 1970s and 1980s sparked heated debate among feminist activists. For example, Barbara Berg (1995) discusses how infertile women are often left out of feminist support for motherhood and are instead sometimes chastised for wanting the "natural" experience of pregnancy but going about it in an "artificial" way. For Berg (1995, 85), "To dismiss [infertile women's] desire to experience pregnancy, childbirth, and rearing their own biologically related children is to endorse traditional patriarchal symbols of achievement". Berg additionally notes that others argue a "real" feminist would not desire motherhood via assisted reproduction because they argue that consciousness-raising is more beneficial to the cause than widening the technological fix. In this case, technology is seen as oppressive. Berg argues against this restrictive feminist model by stating that the use of technology in achieving pregnancy does not reduce the agency of women. Claims stating otherwise are insensitive to infertile women and *their choice* to have children.

New medical frontiers in reproductive medicine will inevitably generate further transformations in understandings of kinship and family. Artificial reproductive technologies (ARTs) are opening up reproductive options for those who cannot conceive "naturally", leading to further questions over who does and who does not have the "right" to bear children. To illustrate this point, we discuss three new (and controversial) technological advances in reproductive medicine that are currently changing the landscape of reproduction: mitochondrial transfer technique, uterine transplants and ovarian tissue harvesting.

Mitochondrial transfer technique takes the DNA from the egg of a mother whose own eggs have damaged mitochondria, and places it in a donor egg with healthy mitochondria before fertilisation in order to avoid passing the mother's mitochondrial condition to the child. The mother is biologically *and* genetically the parent because no nuclear DNA from the donor egg is involved, yet she requires donated tissue from another woman to have a healthy baby. The technique has therefore been colloquially called "three parent" IVF, highlighting concerns that this technique operates contrary to "normal" two-person routes to conception. It has been approved in some countries for women with mitochondrial disease (such as the U.K.) and led to the first birth of a healthy female baby in the Ukraine in 2017 (Scutti 2017).

Uterine transplants are another newly emerging technology that are rewriting the reproductive field. Uterine factor infertility affects as many as 5 per cent of women worldwide (Goldschmidt 2016); it is believed that a transplanted uterus may allow many of these women to have a healthy pregnancy. In practice globally since 2000, attempts at a pregnancy had been unsuccessful until October 2014 when Sweden reported the first birth from a woman with a uterine transplant. The second (and only subsequent birth as of May

2018) was in the United States in November 2017. The success of uterine transplants raises the spectre of the next frontier in treating infertility: the development of artificial wombs, a controversial touchstone for feminist debates over reproductive technology since the 1970s.

One further recent development is ovarian tissue harvesting, which aids women after chemotherapy. This procedure is currently only done at one clinic in the United States. A slice of ovarian tissue containing eggs is harvested and frozen before chemotherapy treatments and then transplanted back into the ovary after treatment to restore reproductive function. Unlike this experimental technique, oocyte or mature egg harvesting is a well-established method of obtaining eggs for use in infertility treatment. More controversial is the emergence of "social" egg freezing, in which healthy women opt to store some of their eggs for future use. Like ovarian tissue harvesting, egg banking is seen to offer the promise of extending fertility into the future.

All of these examples illustrate how the spatial and temporal dimensions of fertility, pregnancy and birth are undergoing significant transformations. These new technologies have the potential to extend understandings of reproduction, fertility and kinship in new directions, made possible by the mobility of bodily tissues and the expansion of who is able to conceive. For critics, these processes are inappropriate technological interference in natural processes. For supporters, ARTs provide ways to address fertility inequalities. Jurgen De Wispelaere and Daniel Weinstock (2014) discuss state assistance for ARTs for infertile citizens arguing for "positive rights" to reproduce, which should involve financial subsidisation by the state. As Barbara K. Rothman has argued, "The treatment of infertility needs to be recognized as an issue of self-determination. It is as important an issue for women as access to contraception and abortion, and freedom from forced sterilization" (quoted in Berg 1995, 103). In addition to Rothman's emphasis on self-determination, the concept of "stratified reproduction" explores the social and cultural barriers to reproduction and the power relations and inequalities embedded in ARTs—specifically, the empowerment of those who are seen as better able to nurture and reproduce due to marital status, sexuality or affluence, in contrast to those who are disempowered (Culley et al. 2009).

Geographical approaches to reproduction

As this discussion shows, reproductive justice movements and new reproductive technologies are raising important questions about the changing social and spatial dimensions of reproductive life. Geographers have a long history of exploring reproductive phenomena. Traditionally, however, much of this work has occurred within population geography, focusing particularly on the implications of fertility patterns for wider patterns of population growth and decline. Geographers have been critical to helping shape the "demographic imagination" in which birth, life and death became objects of scientific and political concern. James Tyner (2013, 702) argues that population geographers need to further consider the politics of fertility as well as mortality and mobility to move from simple demographic questions towards a critical approach that asks: "Within any given place, who lives, who dies, and who decides?" A focus on reproductive geographies encourages us to apply this question to the start of life as much as it has historically been focused on its end.

The last three decades of critical and feminist geographical work has diversified approaches to reproductive issues, with a burgeoning interest in the experiential and political dimensions of reproductive processes. These works argue for expanding geographical accounts of reproduction to a wider array of experiences and subjects, as well as to recognising the politicised and politicising dimensions of reproductive life. Phenomenological and socio-cultural accounts of the embodied experience of the body's fluidity in pregnancy and birth have been developed by feminist geographers such as Robyn Longhurst in a series of works (Longhurst 2001, 2008; see also Teather 1999). Longhurst's *Maternities: Gender, bodies, and space* (2008) focuses on the spaces of maternity and mothering, including body image, breastfeeding, sexuality and the role of digital media and other cultural representations in shaping embodied experiences of mothers. Longhurst's work is a notable geographical contribution to the interdisciplinary field of "motherhood studies", in which maternity is understood as an important psycho-social, cultural and philosophical phenomenon (Baraitser 2008; Lintott and Sander-Staudt 2012; Adams and Lundquist 2013). Geographers are also extending critiques of the medicalisation of reproduction to explore fertility, pregnancy and birth in new ways (Dyck et al. 2001; Klimpel and Whitson 2016). These ideas form an important theoretical starting point for the collection developed here.

Extending the study of reproduction to the broader social practices of sustaining life, feminist economic geographers have long considered maternity/motherhood as part of a political economy of caring labour. The extensive literature on social reproduction in economic geography tends to foreground a much broader range of practices supporting and sustaining daily life. While the concept of social reproduction encompasses the daily routines of parenting and caring for dependent others, work in this area often glosses over the bodily specificities of fertility, pregnancy and birth, leading the editors of a recent collection on social reproduction to call for more attention by feminist economic geographers to "new materialist feminist" approaches to the body (Meehan and Strauss 2015).

Anthropological and sociological accounts of reproduction also signal the importance of social and economic forces such as consumption, neoliberalism and migration on the experience of becoming pregnant, giving birth or seeking fertility therapy (Taylor et al. 2004; Waldby and Mitchell 2006; Gill and Scharff 2011; Cooper and Waldby 2014; O'Donohoe et al. 2014). Despite cultural connotations of selfless giving and generosity, motherhood is also deeply intertwined with practices of consumption in capitalist societies. This dimension of motherhood has been explored through studies of foster care, transnational adoption, midwifery, prenatal healthcare and homebirth (Taylor et al. 2004). In examinations of the spatial dimensions of mothering, researchers note how consumer capitalism plays a large part in constituting maternal subjects, from the relationship of reproductive "rights" with the notion of ownership over one's body, to the overlapping and often entangled notions of economic and affective value attached to child-rearing (Taylor et al. 2004; Thomson and Sørensen 2006).

There is significant literature on the profound transformations made possible by reproductive technologies in reconfiguring kinship and family forms (Strathern 1992; Ginsburg and Rapp 1995; Thompson 2005; Franklin 2013; Nahman 2013). Geographers' attention is now also turning to reproductive and genetic technologies and the social, political and economic transformations to fertility, pregnancy and birth they make possible (Nash 2015; Schurr and Friedrich 2015; Lewis 2016). Geographers have begun to consider how transnational movements of medical personnel, patients and organs have the potential to extend geographical attention to the body's interior as well as to the national, institutional and domestic spaces through which bodily fertility, pregnancy and birth are experienced and take place (Parry et al. 2015). The mobility and mutability of "flesh on the move" across boundaries and territories poses a fundamental challenge to the presumptions of human sovereignty and wholeness that underpin geopolitics (Dixon 2015).

Finally, by analysing reproduction in relation to the exercise of state and governmental power, feminist scholars, and increasingly geographers, are exploring what Penelope Deutscher (2010) refers to as the "procreative hinge" in theories of biopolitics developed by Michel Foucault. Deutscher writes that theorists of biopolitics should be more concerned with reproduction, not as a "niche" effect of population-based politics, but as central to the life-maximising and life-negating dimensions of power. Reproduction is often overlooked by theorists of biopolitics more interested in the grand politics of states, security apparatuses and figures of political "Man". Recent geographical research on abortion (Freeman 2017), public health (Mansfield 2012; Sziarto 2017) and surrogacy (Lewis 2017; Schurr 2017) explores how the biopolitics of reproduction is also central to racialisation, to environmental politics and to imagining new relationships to technology.

Reproductive geographies: Bodies, place, politics

This book aims to bring together emerging geographical work in these areas and to set out a research agenda for future work on what we are calling "reproductive geographies". Taken as a whole, this book brings new empirical, methodological and conceptual perspectives to bear on analyses of reproduction, affirming how the sites, practices and experiences of fertility, pregnancy and birth are central to understanding key geographical concerns, from citizenship to health. This project is distinctly geographical through its exploration of different scales (body, home, local, national, global), boundaries (interior and exterior, public and intimate) and places (of the body and the landscape). It brings original research in significant areas of geographical inquiry such as the spatial dimensions of fluid embodiment, pregnancy and birth together with an up-to-date exploration of the changing landscapes of reproduction, including the expansion of reproductive technologies such as surrogacy and intrauterine insemination.

Acknowledging that reproductive issues are often highly medicalised in academic and popular culture, we seek here instead to use a socio-cultural lens to explore reproductive experiences beyond the clinical. Fertility, pregnancy and birth are embodied, cultural and political processes where the physical and the social intertwine. At the same time, we take up the call by feminist new materialist scholars to re-engage with the biological body by bringing a range of phenomena together around the concept of reproduction and in particular bodily processes that do not necessarily result in parenthood (Alaimo and Hekman 2008; Pitts-Taylor 2016). Feminist scholars have extensively critiqued the extent to which women as subjects are overly identified with reproduction or maternity. This volume confirms the extent to which the bodies and experiences of women are the focus of efforts to control and manage reproductive processes. We include physiological experiences such as the embodied experiences of menopause, the procreative body (encompassing conception, pregnancy and childbirth) and finally limitations of reproductive bodies (including infertility issues and pregnancy termination). Taken together, these topics explore the lived experiences of women with respect to reproductive capacities, overtly acknowledging that for many women their experiences of these issues may not actually be focused on having children. As such, reproductive geographies seep into the everyday lived experiences of people throughout their lives, not just in the limited period where they are seeking to achieve a pregnancy or, for that matter, to avoid one.

A common theme underlying the chapters is how advances in medicine and socio-political changes affect people's experiences of fertility, pregnancy and birth in different ways. Our collection includes dimensions of maternity that are not explicitly addressed by previous work on social reproduction, maternal bodies or reproductive health (see work by Woliver 2002; Taylor et al. 2004; Longhurst 2008; Culley et al. 2009; Nash 2014; McNiven 2016), as we include experiences of fertility technology, surrogacy and biobanking. We extend geographical attention to maternity to new sites as well as to dimensions of embodiment that are not directly about maternity, including menopause and the mobility of bodily tissues such as sperm. International case studies are an important contribution of this project, with chapters focused on specific aspects of reproductive geographies in China, the former Yugoslavia, the U.S., Ghana, the U.K., India and Japan.

Structure of the book

We use three key geographic frames to organise this collection of essays: *bodies, places* and *politics*. Each of the thematic sections draws out significant conceptual strands of geographical scholarship on reproduction, including biopolitics, political ecology, spatial boundaries and mobilities. Taken together, the chapters in this book highlight how scholarship on fertility, pregnancy and birth is taking the biological body into consideration without viewing reproductive issues as simply biological. This emphasis on both the biological *and* the more-than-biological reflects wider efforts by contemporary feminist scholars to revisit bodily processes in a way that avoids reducing them to essentialist understandings of the body as fixed or determined by biology (Colls 2012).

Part I explores different dimensions of how the reproductive *body* and *bodily materials* figure in wider understandings of science, family-making and development. The first two chapters in this section explore how embryos are transformed into scientific objects. In chapter one, Maria Fannin examines how the liberalisation of abortion laws in post-war Japan helped facilitate the development of the largest collection of embryos in the world, the Kyoto Collection. Her chapter calls for greater attention to the *material* geographies of collecting human body parts in modern science. Juliane Collard's chapter brings the collecting of embryos up to the present day, exploring how IVF clinics define the abnormal embryo and in so doing transform "surplus" bodily materials into valuable research tools. Human genes and chromosomes are now able to be visualised and mapped and, as such, are a geographical concern.

Marcia England's chapter also considers the mobility of body parts outside and beyond the "body proper", tracing the spatial trajectories of sperm used in fertility clinics and do-it-yourself efforts to conceive. She details the efforts devoted to choosing a sperm donor, deciding where to inseminate and the need for greater consideration of the complex geographies and "choreographies" (Thompson 2005) involved in becoming pregnant. Place is critical to women's experience of artificial insemination in terms of the location at which the procedure is performed, with some preferring the intimate, private space of the home and others a more "clinical" experience.

Amita Bhakta, Julie Fisher and Brian Reed conclude Part I's emphasis on bodily aspects of reproduction by shifting attention to the bodily dimensions of the perimenopause for women in Ghana. They argue that the specific needs of perimenopausal women, and the way that they perceive the embodied experience of the menopause, have significant implications for their hygiene and sanitation needs and for the material infrastructures needed to accommodate the menopause.

This part of the book underscores how the body, including the body's interior and bodily materials that circulate outside the body, is an important site for geographical analysis. In this section, pregnancy is both a sought-after and often elusive achievement on the part of women and their partners involved in fertility therapy in the discussions of Collard and England, and a phenomenon brought to an early end and then made the object of scientific study in Fannin's chapter on embryo collecting. Menstruation and the menopause should not be understood solely through their relationship to fertility; they also raise questions about the presumptions surrounding embodiment built into the infrastructural environment, as Bhakta, Fisher and Reed make clear. Engaging in different ways with notions of porosity, boundary-making and mobility, these chapters illustrate how attention to the space or scale of the body is essential to understanding reproductive processes.

In Part II, we emphasise how the *places* and *spaces* of reproductive activity have strong symbolic meaning. Cultural understandings of place can profoundly influence the lived experiences of individuals at critical points in their reproductive lives. Katie Merkle's chapter provides a link between ideas of place and embodiment by arguing that the lived experiences of pregnant graduate students are strongly influenced by the particularities of academic institutions and the hierarchies of student/supervisor relationships. Here, place is seen as constraining the continued daily activities of pregnant people at a particular point in their academic lives. In Chapter 6, Dalia Bhattacharjee examines how commercial surrogacy in India operates in a "third space" that is neither public nor private, in which women working in the surrogacy industry are not fully considered as workers, despite the nature of the work they do as surrogate mothers. Her chapter discusses the reproductive "labour" required in this job, dismissing the notion that surrogates are "replaced" mothers. She goes on to argue that place-based restrictions on reproductive labourers leave them effectively under surveillance for the period of their pregnancy, highlighting power imbalances embedded in the surrogacy industry.

Risa Whitson and Helen Hazen both consider the place-based preferences of women who consciously choose an out-of-hospital birth for their babies. Whitson considers why place is so significant to the birth process that some women will deliberately seek an "alternative" birth location outside the hospital, despite the huge societal pressures that promote hospital birth. She discusses how the women in her study experienced their homes as spaces of resistance to medical norms, enabling more meaningful connections to place. Helen Hazen's chapter on birth centre births in Minnesota extends this discussion of the diversity of meanings associated with giving birth outside of hospital by looking at a growing interest in birth centres and "in-between" spaces that are touted as providing the safety of hospital with the comfort of home. In her chapter, the alternative space of the freestanding birth centre figures as a compromise weighed between different notions of risk and responsibility on the part of birthing women, their families and the institutions of home and hospital.

As a whole, Part II suggests that attention to place is critical to understanding the changing territories of people's reproductive lives. This occurs both as people's bodies and physical capabilities change across their lifetimes, and as broader societal understandings of the reproductive body are brought to bear on places of pregnancy and birth. In these chapters, pregnancy is characterised by myriad forms and sites of surveillance: medical, institutional, academic. The hyper-surveillance of pregnant surrogates and the efforts of out-of-hospital birthers to escape the medical gaze, if only partially and temporarily, underscore the extent and depth of the exercise of power to preserve life, and the condensation of this power around the pregnant and birthing body.

In Part III, we focus on *the politics of reproduction* as a key way in which space is navigated and renegotiated. Two chapters in this part of the book bring questions of reproduction to bear on state policies of belonging and citizenship. They highlight what Penelope Deutscher (2010) insists is the centrality of reproduction to the broader dynamics of biopolitics. In his chapter, Robert Kaiser focuses on the public discourse surrounding pregnant mainland Chinese women giving birth in Hong Kong, seeing an example of changing modes of biopolitical power in the shifting policies around maternity migration. He demonstrates how population policies, concerns over the allocation of public resources and anti-immigrant sentiments coalesce on the bodies of women at territorial borders. On the border, the "signs" of pregnancy (e.g., a swollen abdomen) signify the intention to circumvent uneven rights of citizenship between Hong Kong and mainland China. Carl T. Dahlman's chapter considers how the former Yugoslav republics targeted women's bodies as key strategic sites for the application of biopolitical concerns over ethnic and national power. His historical narrative and focus on fertility and nationalism demonstrate the breadth of reproductive concerns shaping nation-building in the aftermath of violent political conflict. Concluding this section, Andrea Rishworth and Jenna Dixon shift attention to the spaces of global health policy. They bring insights from feminist political ecology to bear on research carried out in northern Ghana, arguing for the importance of connecting maternal and child health to the health of the wider environment.

As a whole, Part III highlights the spatial elements of fertility, birth and maternal and child health through the lens of geo- and ecological politics. The body, the state and the globe all intertwine in this section to illuminate the political aspects of reproduction. All three chapters in this section engage with how global systems, whether of nation-states or supra-national health institutions, are intimately concerned with women's bodies as reproductive and with making women's reproductive capacities the focus of national and international agendas.

Reproductive Geographies brings together diverse approaches to reproductive bodies, spaces and politics situated across different geographical subfields, from geopolitics to global public health. Our aim for this volume is not to develop one single methodological or conceptual framework for the study of reproductive processes but rather to bring together previously disparate strands of research on reproduction, broadly conceived, that may only rarely speak to each other. Each part of the book explores different conceptual currents in theorising the geographies of reproduction: attentiveness to multiple dimensions of reproductive bodies and body parts in Part I, the situated nature of fertility, pregnancy and birth as spatial processes in Part II and the centrality of reproduction for bio-, geo- and ecological politics in Part III. As the chapters in this book demonstrate, reproductive bodies, places and politics are central to debates over citizenship, justice, care and technology. We offer this collection of essays as a call for sustained *geographical* exploration of the wider economies and ecologies of reproduction today.

References

Abel, Sally, and Robin A. Kearns. 1991. "Birth Places: A Geographical Perspective on Planned Home Birth in New Zealand." *Social Science & Medicine* 33(7): 825–834.

Alaimo, Stacy, and Susan J. Hekman. (Eds.) 2008. *Material Feminisms*. Indianapolis: Indiana University Press.

Adams, Sarah LaChance, and Caroline R. Lundquist. (Eds.). 2013. *Coming to Life: Philosophies of Pregnancy, Childbirth and Mothering*. New York: Fordham University Press.

Baraitser, Lisa. 2008. *Maternal Encounters: The Ethics of Interruption*. London: Routledge.

Bashford, Alison. 2014. *Global Population: History, Geopolitics and Life on Earth*. New York: Columbia University Press.

Berg, Barbara J. 1995. "Listening to the Voices of the Infertile." In *Reproduction, Ethics, and the Law: Feminist Perspectives*, edited by Joan C. Callahan, 80–108. Indianapolis: University of Indiana Press.

Boyer, Kate. 2012. "Affect, Corporeality and the Limits of Belonging: Breastfeeding in Public in the Contemporary UK." *Health & Place* 18(3): 552–560.

Boyer, Kate. 2018. *Spaces and Politics of Motherhood*. Lanham, MD: Rowman & Littlefield.

Colls, Rachel. 2012. "Feminism, Bodily Difference and Non-representational Geographies." *Transactions of the Institute of British Geographers* 37(3): 430–445.

Cooper, Melinda, and Catherine Waldby. 2014. *Clinical Labour: Human Research Subjects and Tissue Donors in the Global Bioeconomy*. Durham, NC: Duke University Press.

Culley, Lorraine, Nicky Hudson, and Floor Van Rooij. 2009. "Introduction: Ethnicity, Infertility, and Assisted Reproductive Technologies." In *Marginalized Reproduction: Ethnicity, Infertility and Reproductive Technologies*, edited by Lorraine Culley, Nicky Hudson, and Floor Van Roojj, 1–14. London and Sterling, VA: Earthscan.

De Wispelaere, Jurgen and Daniel Weinstock. 2014. "State regulation and assisted reproduction: Balancing the interests of parents and children" in *Family-Making: Contemporary Ethical Challenges*, edited by Francoise Baylis and Carolyn McLeod, 131–150. Oxford: Oxford University Press.

Deutscher, Penelope. 2010. "Reproductive Politics, Biopolitics and Auto-immunity: From Foucault to Esposito" *Journal of Bioethical Inquiry* 7(2): 217–226.

Dixon, Deborah P. 2015. *Feminist Geopolitics: Material States*. Farnham: Ashgate.

Douglas, Mary. 1966. *Purity and Danger*. London: Ark Books.

Dyck, Isabel., Nancy Davis Lewis and Sara McLafferty (Eds.). 2001. *Geographies of Women's Health: Place, Diversity and Difference*. Abingdon: Routledge.

Edelman, Lee. 2004. *No Future: Queer Theory and the Death Drive*. Durham: Duke University Press.

Fannin, Maria. 2003. "Domesticating Birth in the Hospital: 'Family-centered' Birth and the Emergence of 'Homelike' Birthing Rooms." *Antipode* 35(3): 513–535.

Franklin, Sarah. 2013. *Biological Relatives: IVF, Stem Cells, and the Future of Kinship*. Durham: Duke University Press.

Freeman, Cordelia. 2017. "The Crime of Choice: Abortion Border Crossings from Chile to Peru" *Gender, Place & Culture* 24(6): 851–868.

Gill, Rosalind and Cristina Scharff (Eds.) 2011. *New Femininities: Postfeminism, Neoliberalism and Identity*. Basingstoke: Palgrave Macmillan.

Ginsburg, Faye and Rayna Rapp (Eds.) 1995. *Conceiving the New World Order: The Global Politics of Reproduction.* Berkeley: University of California Press.

Goldschmidt, Debra. 2016. "Living Donor Uterus Transplants Performed in US." CNN. October 5. https://www.cnn.com/2016/10/05/health/us-uterus-transplants/index.html

Griffin, Jo. 2018. Brazilian Women Braced for Battle Amid Simmering Fears over Abortion. *The Guardian,* April 28. https://www.theguardian.com/global-development/2018/apr/26/brazil-women-braced-for-battle-simmering-fears-abortion-law

Klimpel, Jill and Risa Whitson. 2016. "Birthing Modernity: Spatial Discourses of Cesarean Birth in São Paulo, Brazil." *Gender, Place & Culture 23(8)*: 1207–1220.

Kornelsen, Jude, Andrew Kotaska, Pauline Waterfall, Louisa Willie, and Dawn Wilson. 2010. "The Geography of Belonging: The Experience of Birthing at Home for First Nations women." *Health & Place 16(4)*: 638–645.

Lane, Rebecca. 2013. "Healthy Discretion? Breastfeeding and the Mutual Maintenance of Motherhood and Public Space" *Gender, Place & Culture 21(2)*: 195–210.

Lewis, Sophie, 2016. "Gestational Labors: Care Politics and Surrogates' Struggle." In *Intimate Economies: Bodies, Emotions, and Sexualities on the Global Market,* edited by Susanne Hofmann and Adi Moreno, 187–212. New York: Palgrave Macmillan.

Lewis, Sophie, 2017. "Defending Intimacy against What? Limits of Antisurrogacy Feminisms." *Signs: Journal of Women in Culture and Society 43(1)*: 97–125.

Light, Alexis D.; Obedin-Maliver, Juno; Sevelius, Jae M.; and Jennifer L. Kerns. 2014. "Transgender Men who Experienced Pregnancy after Female-to-male Gender Transitioning." *Obstetrics & Gynecology 124(6)*: 1120–1127.

Lintott, Sheila and Maureen Sander-Staudt (Eds.) 2012. *Philosophical Inquiries into Pregnancy, Childbirth and Mothering.* London: Routledge.

Longhurst, Robyn. 2001. *Bodies: Exploring Fluid Boundaries.* London: Routledge.

Longhurst, Robyn. 2008. *Maternities: Gender, Bodies and Space.* London: Routledge.

Mansfield, Becky. 2012 "Gendered Biopolitics of Public Health: Regulation and Discipline in Seafood Consumption Advisories." *Environment and Planning D: Society and Space 30(4)*: 588–602.

McDowell, Linda. 1999. *Gender, Identity and Place: Understanding Feminist Geographies.* Minneapolis: University of Minnesota Press.

McNiven, Abi. 2016. "'Geographies of Dying and Death' in Relation to Pregnancy Losses: Ultrasonography Experiences." *Social & Cultural Geography 17(2)*: 233–246.

Meehan, Katie, and Kendra Strauss. (Eds.). 2015. *Precarious Worlds: Contested Geographies of Social Reproduction.* Athens, GA: University of Georgia Press.

Nahman, Michal. 2013. *Extractions: An Ethnography of Reproductive Tourism.* New York: Palgrave Macmillan.

Nash, Catherine. 2015. *Genetic Geographies: The Trouble with Ancestry.* Minneapolis: University of Minnesota Press.

Nash, Meredith (Ed.). 2014. *Reframing Reproduction: Conceiving Gendered Experiences.* London: Palgrave MacMillan.

O'Donohoe, Stephanie, Pauline Maclaran, Margaret Hogg, Lydia Martens, and Lorna Stevens. (Eds.). 2014. *Motherhood Markets and Consumption: The Making of Mothers in Contemporary Western Culture.* Abingdon: Routledge.

Parry, Bronwyn, Greenhough, Beth, Brown, Tim, and Isabel Dyck (Eds.). 2015. *Bodies across Borders: The Global Circulation of Body Parts, Medical Tourists and Professionals.* London: Ashgate.

Pitts-Taylor, Victoria (Ed.) 2016. *Mattering: Feminism, Science, and Materialism.* New York: NYU Press.

14   *Fannin, Hazen and England*

Roberts, Dorothy. 1997. *Killing the Black Body: Race, Reproduction, and the Meaning of Liberty*. New York: Pantheon Books.
Rose, Gillian. 1993. *Feminism and Geography: The Limits of Geographical Knowledge*. Minneapolis: University of Minnesota Press.
Schurr, Carolin. 2017. "From Biopolitics to Bioeconomies: The ART of (Re-) producing White Futures in Mexico's Surrogacy Market." *Environment and Planning D: Society and Space*, 35(2): 241–262.
Schurr, Carolin and Bettina Friedrich, 2015. "Serving the Transnational Surrogacy Market as a Development Strategy?" In *The Routledge Handbook of Gender and Development*, edited by Anne Coles, Leslie Gray and Janet Momsen, 236–243. London: Routledge.
Scutti, Susan. 2017. "First Baby Girl Born Using Controversial IVF Technique." CNN. January 18, 2017. https://www.cnn.com/2017/01/18/health/ivf-three-parent-baby-girl-ukraine-bn/index.html.
Silliman, Jael, Marlene Gerber Fried, Loretta Ross and Elena R. Gutiérrez. 2004. *Undivided Rights: Women of Color Organize for Reproductive Justice*. Cambridge, MA: South End Press.
Somashekhar Sandy and Amy B. Wang. 2017. "Lawmaker Who Called Pregnant Women a 'Host' Pushes Bill Requiring Fathers to Approve Abortion". *The Washington Post*. https://www.washingtonpost.com/news/post-nation/wp/2017/02/14/oklahoma-bill-would-require-father-of-fetus-to-approve-abortion/?noredirect=on&utm_term=.499b6a746810
Strathern, Marilyn. 1992. *Reproducing the Future: Anthropology, Kinship, and the New Reproductive Technologies*. London: Routledge.
Sziarto, Kristin M. 2017. "Whose Reproductive Futures? Race-biopolitics and Resistance in the Black Infant Mortality Reduction Campaigns in Milwaukee." *Environment and Planning D: Society and Space* 35(2): 299–318.
Taylor, Janelle S., Layne, Linda L., and Danielle. F. Wozniak (Eds.). 2004. *Consuming Motherhood*. New Brunswick: Rutgers University Press.
Teather, Elizabeth Kenworthy (Ed.). 1999. *Embodied Geographies*. London: Routledge.
The Endocrine Society. 2018. "Dimethandrolone Undecanoate Shows Promise as a Male Birth Control Pill." *ScienceDaily*. www.sciencedaily.com/releases/2018/03/180318144834.htm (accessed June 7, 2018).
Thomsen, Thyra Uth and Sørensen, Elin Brandi. 2006. "The First Four-wheeled Status Symbol: Pram Consumption as a Vehicle for the Construction of Motherhood Identity", *Journal of Marketing Management*, 22(9-10): 907–927.
Thompson, Charis. 2005. *Making Parents: The Ontological Choreography of Reproductive Technologies*. Cambridge, MA: MIT Press.
Tyner, James A., 2013. "Population Geography I: Surplus Populations". *Progress in Human Geography, 37(5)*: 701–711.
Waldby, Catherine. and Robert Mitchell. 2006. *Tissue Economies: Blood, Organs and Cell Lines in Late Capitalism*. Durham: Duke University Press.
Woliver, Laura. 2002. *The Political Geographies of Pregnancy*. Champaign, IL: The University of Illinois Press.

Part I
Bodies

1 Making an "embryological vision of the world"

Law, maternity and the Kyoto Collection

Maria Fannin

Lynn Morgan's 2009 account of American embryo collecting, *Icons of Life*, describes the creation of embryo collections in the U.S. and Europe over the course of the late nineteenth and early twentieth centuries. Embryos were obtained by doctors, most often from women who had miscarried a pregnancy or whose pregnancies were discovered after hysterectomy. Preserved in formalin or other chemical solutions and prepared for examination, the specimens provided teaching and research resources to study embryo morphology and development over time.[1] These collections, Morgan argues, were essential to the establishment of embryology as a science. Embryo collections informed the medical study of pregnancy but also helped shape the development of other sciences, from evolutionary biology to genetics. Collections of embryonic material could be seen as continuous with the practices of anatomists from an earlier age who sought to collect and preserve specimens of humans and animals—and their body parts—for display and study. Embryo collections also sit at the cusp of the genetic and molecular sciences that would soon begin to dominate the study of human development. As Morgan writes, embryological collections inaugurated what would become the central role played by biology—surpassing chemistry—in late twentieth-century science. The embryological collections of the nineteenth and twentieth centuries sought to probe the "genesis" of life and its variability and could be said to anticipate the molecular and genetic sciences to come.

This chapter focuses on the social and spatial history of the Kyoto Collection, a collection of embryonic material established in 1961 by Dr Hideo Nishimura at Kyoto University in Kyoto, Japan. In doing so, I extend the insights of Morgan's influential social history of the making of embryo collections by scientists in the U.S. and Europe to their counterparts in Japan. Like other embryo collections, the Kyoto Collection, as it is now known, was part of twentieth-century efforts to systematically study human development from conception to birth. It also aimed to illustrate variation across embryos deemed to be normal as well as the presence of developmental "anomalies". Much of the material in the collection originated from elective termination of pregnancies, made possible by the legalisation of abortion in Japan under the 1948 Eugenics Protection Law. The collection now houses over 44,000 human embryos and foetuses and is considered the largest human embryo collection in the world. Japan's post-war history of eugenic policy, innovations in the medical surveillance of pregnancy and abortion practice helped shape the development and reception of the collection.

Geographical accounts of archival and museum collections emphasise how relations between collectors and objects, and the collections themselves, reveal the spatial and historical preoccupations of curators and collectors (Hill 2006). At its inception, the Kyoto Collection differed from previous embryo collections. Whereas Morgan describes the collection of embryos in American embryological collections as somewhat ad-hoc and opportunistic, drawing on the goodwill of specific doctors through a relatively small network to build up a collection, Nishimura was able to enrol a large number of doctors who would regularly send materials to Kyoto accompanied by demographic and behavioural data about the women from whom the material had originated, and they carried out these tasks over a considerable period of time. The acquisition of material for the Kyoto Collection continued well into the latter half of the twentieth century. At a time when most other embryological collections in Europe and the U.S. had long ceased actively collecting specimens, the Kyoto Collection continued.

The number of embryos collected by Nishimura thus greatly exceeded those of other embryo collectors. The embryos in the Kyoto Collection were brought together in what one researcher described as a "random manner" by numerous doctors from around Japan. Physicians sending specimens were not asked to select specimens based on specific criteria (for example, whether the foetus was alive or dead when removed from the pregnant woman's body or whether it was visibly normal or abnormal). Because of this, the Kyoto Collection is considered by researchers to be "representative of the total intrauterine population in Japan" (Nagai et al. 2016, 112). This argument is based on the view that embryos collected primarily from miscarriage were more likely to be abnormal (and thus unviable); those collected primarily through elective termination, like the embryos in the Kyoto Collection, would represent an "unbiased intrauterine population", a quality that renders the collection valuable for population-based study (Kameda et al. 2012, 48). Together, the collection of biological materials and behavioural data was informed by a kind of "epidemiological reason" (Reubi 2018) that shaped how collectors and curators of the study envisioned their work as leading to a better understanding of the health and development of the Japanese population.

The study of population through the application of epidemiological reason is often told in histories of medical geography through the iconic figures of John Snow or William Budd, rather than through the efforts of anatomists and tissue collectors. Yet the collectors and curators of preserved human tissue aimed to glean the truth of the body's health through the mapping, classification, comparison and analysis of the body's interior. These practices often drew analogies between territory and body and sought to understand how the body was related to and influenced by its environment, including the intrauterine environment. In this way, anatomical study of the body was also a kind of cartographic practice, generating a spatial conception of the body and its relation to other forces within, outside and across the body's supposed boundaries of interior and exterior. In linking the collection of embryos to the broader social practices surrounding the body and the political and legal spaces of abortion, studies of human tissue collections also demonstrate how parts of bodies were made into scientific objects. As Morgan (2004, 4) writes,

> Transforming women's calamities into embryo specimens was a cultural achievement that was possible only because most people attached little (if any) moral importance to dead human embryos. In the early 1900s, non-viable human embryos were valued only by embryologists, which made it relatively easy to render them anonymous and a-social.

The creation of embryo collections can reveal the changing moral value of bodies and body parts. They can also reveal changing geographies of reproduction as the end of a pregnancy, whether intended or not, moves into the space of the clinic and is "medicalised". Medical and anatomical collections and their contemporary counterparts, tissue and cell biobanks, are thus important resources for understanding the history and geography of medicine, and for exploring the bodily geographies of contemporary science and technology.

Morgan's history of embryo collections reveals how these collections became potent visual resources in struggles over reproductive politics in the United States. Inspired by Morgan's work, this chapter asks why and how the Kyoto Collection began actively collecting specimens—and continued to collect them—long after many other embryological collections in the U.S. and Europe had ceased collecting embryos. What social, economic and regulatory conditions made embryo collecting possible in Japan? And how is this collection of preserved embryos regarded today? Embryo collections, and collections of human biological materials in general, are critical, I argue, to understanding the reproductive geographies of contemporary science.

Creating the Kyoto Collection

Hideo Nishimura's early research focused on the anatomy of bullfrogs and was influenced by the work of Friedrich Kraus, an Austrian internist whose book *General and Special Pathology of the Individual* (1926) provided inspiration for engaging in "studies of development as the basis of human life" (Tanimura 1996, 3). Accounts of Nishimura's career during the 1940s emphasise how the war made accessing scientific journals published abroad extremely difficult; after the war Nishimura's work shifted to experimental embryology, where he began to focus his attention on the "intrauterine environment". His interest at the time was on the effects of exposure to chemicals such as nicotine, caffeine and aminoazobenzene derivatives (then used in the treatment of bacterial infections and as a bright yellow dye) on the development of mammalian embryos and foetuses. By 1960, his research assistant Takashi Tanimura recounts that his work had become increasingly concerned with developmental processes in the human embryo and foetus, influenced by the emerging research from the U.S. and Europe on embryo development.

Nishimura was also, Tanimura suggests, influenced by reports of the effects of thalidomide on developing embryos. Thalidomide was sold in Japan until 1962, nearly a year after the drug had been removed from the market in some other countries (Lenz 1988). The global thalidomide disaster, in which pregnant women were prescribed medication to treat nausea that was later found to cause serious problems in the developing foetus, including missing or malformed limbs, was identified as a turning point in Nishimura's interest in developmental processes. Nishimura's professional activities at this time also included a key role in the establishment of the Japanese Teratology Society, where teratology is defined as the study of abnormal development. In 1961, Nishimura visited the U.S. where he intensified his interest in the development of the human embryo that would shape the rest of his career. In that same year, he began collecting embryos and foetuses from the termination of pregnancies carried out by doctors around Japan.

Collecting embryos was made possible in part through the coordination of the Japanese Medical Association, which sent invitations to physicians licensed to perform abortions to request their participation. In Nishimura's obituary in 1996, it was noted that his "work was assisted by... more than 200 Japanese medical practitioners" (Nishimura 1996, 1137). But in fact the number of practitioners who contributed to his collection was much larger. Around 1,400 doctors registered initial interest in participating in the development of the collection, and of these over 970 doctors described as highly skilled in obtaining quality specimens sent materials to Nishimura to be retained for future study (Nishimura et al. 1968; Kameda et al. 2012). Nishimura and his team described these physicians as "willing to provide us with better specimens; an arrangement that allowed us to obtain standardized data on normal and abnormal human development during the stages of organogenesis, based on specimens derived from healthy pregnancies" (Nishimura et al. 1968, 281).

Most of the embryos in the collection were from pregnancies terminated in the first trimester. Once the materials were brought to Nishimura's laboratory, they were measured, assessed for developmental "stage" according to a classification schema developed from the Carnegie Collection of embryos (known as the Carnegie Stages), and examined for anomalies. Biographical information about the mother was also sent to Nishimura, including the mother's age, marital status, number of pregnancies, whether she was employed or unemployed, smoked or used alcohol or other drugs including medications, whether the pregnancy ended through "spontaneous or artificial abortion", the mode of delivery and any symptoms of infection or radiation exposure experienced during pregnancy (Kameda et al. 2012, 51). Embryos were sent from 21 prefectures across six districts of the country, and more than 95 per cent of the embryos in the Collection were sent between 1960 and 1979.

The high number of specimens collected was a result of the coordination of physicians and the bodily contributions of pregnant women on a scale much more extensive than that available to other embryo collectors. Nishimura's work was effective in amassing the largest embryological collection in the world in part because of the coordination of a number of Japanese clinicians and the institutional support of both the Japanese Medical Association as well as domestic and overseas funders, including the U.S. National Institutes of Health (Tanimura 1996). Nishimura's ability to call on the medical networks of his contemporaries in clinical practice is evidenced by the scale and coordination of collection.[2] However, the scientific success of collecting was not solely the work of a charismatic individual or a set of dedicated clinicians, but was also shaped by the broader social context of Japanese women's access to abortion during the height of the collection's acquisition of materials.

Eugenic theories in Japan

Eugenic theories circulating in Japan in the latter part of the nineteenth and early part of the twentieth centuries targeted women's bodies as "a strategic site in which constitutional improvement of the Japanese 'race' could be made" (Otsubo 2005, 61). Historically high literacy rates enabled mass media dissemination of eugenic theories and popularised concerns over hygiene, nutrition and "eugenic marriage" as central to race betterment (Robertson 2010). Michiko Suzuki (2013, 42) stresses how attitudes towards population *quality* shifted during the 1930s due to the impact of Japan's entry into war with China in 1937, writing that "eugenic principles to improve 'quality' more or less took a back seat to the need for 'quantity'–that is, population growth". During the 1940s, Suzuki argues, policies and laws, including the 1940 National Eugenics Law (modelled after the 1933 Nazi sterilisation law, itself informed by U.S. laws on sterilisation, see Ogino 1996; Robertson 2010), prohibited sterilisation and abortion for the healthy population. Sterilisation was permitted only for those known to have hereditary diseases.

In the post-war period, however, theories of racial improvement began to move away from articulating the need to increase the population and towards re-emphasising its "quality". These theories were successfully translated into laws in the aftermath of the war that made abortion comparatively accessible. In 1948, the passage of the Eugenic Protection Law made abortion legal under specific circumstances, with commentators attributing this change as a response to the severe food crisis and economic insecurity experienced in Japan at the time (Nishimura et al. 1968, 281). The efforts to rebuild the Japanese economy by limiting population growth, and thus enable the industrialisation and urbanisation of the Japanese workforce, were cited by key legislative proponents of the 1948 law as reasons to expand rather than contract access to abortion. This law made it possible for women to obtain abortion if the pregnancy was the result of rape, if she, her spouse, or a relative had a hereditary physical illness, if she or her spouse had a hereditary *or* non-hereditary mental illness, if either had leprosy or if the pregnancy was likely to endanger her physical health. Earlier efforts to pass a bill giving access to contraception as well as to abortion and sterilisation by reason of "financial hardship" had failed. The successful 1948 Eugenic Protection Law removed the emphasis on birth control and economic criteria. It also authorised only designated doctors to perform abortions and any request for an abortion had to be approved by a "Eugenic Protection Committee".

In 1949, the law was further revised to include a provision permitting abortion for economic reasons. However, women's claims to economic hardship were rarely scrutinised. Miho Ogino (1996, 133) writes,

> The 1948 law declared that induced abortion was legitimate not only for eugenic reasons, rape, or leprosy of the pregnant woman or her spouse, but also "when the continuation of pregnancy or childbirth would be physically detrimental to the health of the mother." In 1949, the phrase "or economically" was added to further extend legitimate reasons for abortion. Since there were no guidelines regarding economic criteria, Japanese women were thus practically given abortion on request.

The termination of a pregnancy for "socio-economic" reasons required the request of the woman and the permission of her spouse or partner, an arrangement that has been described as relatively liberal for its time compared with the restrictions surrounding abortion in most Western countries.

This liberalisation of abortion access, despite restrictions around contraception, was not the result of active campaigning on the part of women's groups—as is reflected in the debates over abortion in Europe and the United States in the 1960s and 1970s—but rather the effect of the ability of Japanese associations of ob-gyns and their leadership in the Japanese parliament to ensure that abortions were provided by doctors, rather than by midwives, other medical professionals or unregulated providers (see Homei 2012). As Samuel Coleman (1992, 18) writes, "abortion became the preserve of a well-organised group of medical specialists who have made it widely available to Japanese married women on a fee-for-service basis". The law required that all abortion providers must be members of prefectural medical associations. These prefectural associations were where collectors solicited materials for the embryo collection. For Nishimura and his colleagues, the legality of abortion in Japan meant fewer obstacles to systematic collection, and indeed the request to collect specimens was carried out at the same prefectural level as the training and regulation of doctors authorised to provide abortions. The coordination of collection was thus facilitated by the structure of the legal regulations around abortion that protected the "market" of ob-gyns for abortion provision.

The Eugenic Protection Law reflected the shifting emphasis within eugenic policy in Japan from promoting population growth in order to support

Japan's imperial ambitions, to post-war efforts to "improve" the population and support modern ideals of "protection of maternal health and life" (Norgren 2001, 41). What the Eugenic Protection Law also made possible was a new kind of "tissue economy" that enabled doctors and researchers to systematically collect embryonic material from healthy women's pregnancies (Waldby and Mitchell 2006).[3] In the U.S. and Europe, many of the specimens sent to embryo collectors in the early twentieth century were obtained from pregnancy terminations carried out either unwittingly, for example upon discover of a pregnancy after hysterectomy (a common treatment in the 1930s for conditions such as uterine fibroids), or under "exceptional" conditions (Morgan 2009). By contrast, abortion was widely accepted as a method of birth control in post-war Japan. Unlike in the U.S. and much of Europe, social and political opposition to abortion in Japan was a relatively minor issue in the post-war period until the 1970s. The systematicity and success of the Kyoto Collection's efforts to procure embryonic and foetal specimens illustrates this social and political reality.

The scale and scope of the Kyoto Collection was shaped by the conditions under which abortion was permitted under Japanese law. It also involved the coordination of hundreds of physicians, and the investments of many hours of time and expertise to prepare embryos for further study. The Collection thus reflects how tissue economies are shaped by the particularities of laws governing life and death, by techniques and practices and by the social and moral significance attached–or not–to embryos. The collection also sheds light on how eugenic policies shaped scientific practice, although not in ways that are self-evident or easily disentangled from other policies aimed at the health of the population. The study of embryological development was not explicitly motivated by eugenic concerns for racial purity or the cultivation of superior populations, but the nineteenth and early twentieth century nascence of embryology shared with more self-consciously eugenic policies the language of "abnormality" and the underlying presumption that human development involved elements of "racial" specificity that required comparative study between populations.

The Kyoto Collection would later be credited with having "greatly contributed to the standardization of the embryonic development of the Japanese, to the study of embryogenesis, and to the analysis of the etiology of congenital malformations" (Fujimoto 2001, 67). The collection is a testament to the coordinated efforts of scientists to create research tools and resources to enable comparison between populations. In this sense, the Kyoto Collection was enrolled in the broader process of *making populations*, linking the collection of biological materials to the study of a population's "birth, death and space" or territory (Bashford 2007, 173). The study and management of population was a central concern of modern states over the course of the nineteenth century. Governing a modern nation-state "encompassed not only an interest in improving and revitalizing populations...but also the obvious, if sinister, corollary that some populations would be unfit to do so" (Levine and

Bashford 2010, 7). Eugenic science was devoted to the "problem" of how best to manage both the quantity and the quality of a population. Embryologists, with their interest in the trajectories of normal and abnormal development, were critical to this enterprise alongside many others.

Making biological value: The Kyoto Collection today

Today, the Kyoto Collection is viewed as a historically significant embryological collection alongside other European and American collections. In an era of big data analysis, the Kyoto Collection's value for researchers is measured by its ability to "stand in" for a population. By contrast, earlier collections such as the Carnegie and Blechschmidt collections include specimens representing specific abnormalities or conversely ideal-types of embryological development; the Carnegie collection was used extensively to develop a model for assessing the developmental age or "stage" of normal embryonic development (now known as the "Carnegie stages"). The sheer volume of embryos in the Kyoto Collection enables researchers to argue for its ongoing value as representative of something altogether different: its scale and the conditions under which embryos were acquired from healthy women means that it is deemed to provide an accurate representation of the "normal" population.

Reporting on a project to generate a digital research database from the analogue collection of papers and punch cards that contain relevant biographical and medical information about the mothers who contributed to the collection, Tomomi Kameda and co-authors (2012) argue that the Kyoto Collection is viewed as amenable to contemporary epidemiological and statistical analysis. The digitisation project has generated a database of over 22,000 embryos and associated data. Researchers involved in the digitisation project suggest that such a large database representing a "random" sample of the population will contribute to the development of a better understanding of the causes of embryonic and foetal anomalies, and so inform the diagnosis of embryonic anomalies and the monitoring of pregnancies during the embryonic period rather than the later foetal stage. The collection now has a digital presence through the Kyoto Human Embryo Visualisation Project (Congenital Anomaly Research Center 2010, see also Hill et al. 2016) and is the subject of studies to develop 3D and 4D reconstructions of embryo development (Yamada et al. 2006). These visualisation techniques are employed to probe the developmental processes of organs in the embryo's interior in an attempt to identify the causes of miscarriage involving embryos that otherwise appear normal (Kanahashi et al. 2016).

Contemporary ethical concerns about access to bodily materials, and specifically embryonic material, also shape the value of the Kyoto Collection today. Compared to the decades spent acquiring biological materials in the Kyoto Collection, "the task of constructing a new collection of human embryos would be both technically and ethically challenging" (Kameda et al. 2012, 53). It would be very difficult to reproduce the collection at a similar scale given current concerns about the ethical use of biological materials and tissues derived from them such as stem cells. The value of the Kyoto Collection is also envisioned through the application of new analytical techniques to the specimens in the collection. Efforts to extract DNA from the wet tissues preserved in formalin and other chemicals are being carried out with the hope, as yet unrealised, that these tissues will "result in a fuller understanding of human congenital anomalies" (Nagai et al. 2016). Genetic research on historical embryo collections poses new ethical challenges, given that the techniques for extracting and analysing DNA were unknown when the materials were being actively collected and donor consent is often impossible to obtain (Asai et al. 2002).

Morgan (2004, 3) situates the social history of embryo collections as a means of tracing the changing moral status of the embryo and foetus over time, writing: "[w]hereas a hundred years ago embryologists were the only ones to care much about the disposition of dead embryos, today they are increasingly viewed as active, animated agents". Embryo collecting today has shifted spatially from the gynaecological clinic to the contemporary spaces of assisted reproduction, where "surplus" embryos are transformed into research resources (see Parry 2006, Waldby and Mitchell 2006, Thompson 2007, see also Collard in this volume). While embryos and foetal materials are still collected from pregnancy termination, the emergence of reproductive technologies that produce excess embryos have transformed the social and moral geographies of embryo and fetal tissue collecting. However, contemporary discussions of embryo donation from IVF treatments highlight the persistence of the invisibility of women's bodily contributions in medical and scientific literatures on donation.[4] As Masae Kato (2014, 252) writes in her account of ongoing research with embryo donors in Japan and their changing conceptions of the "gift" of donation,

> [m]any medical specialists speak as if embryos are just there, like fruits growing on trees (Dickenson, 2007, 60; Tsuge 2002) ... more than 90% of interviewed IVF specialists stated that technological advancement has rendered ova collection painless, and IVF has become so generally accepted that "women do not have special feelings about IVF experience, unlike abortion" (Kato and Sleeboom-Faulkner 2011, 435).

Accounts of the acquisition of materials for the Kyoto Collection maintain that women did provide informed consent for their bodily materials to become part of the collection (Kameda et al. 2012). It is possible that informed consent in this historical context was granted verbally, as it is often today in Japan. There is however no clear evidence of precisely what women understood about their participation in the development of the embryo collection.

The Kyoto Collection bears witness to the efforts of researchers to create an archive of "normal" as well as "pathological" human embryonic development.

But embryo collections reveal more than the history of scientific achievement, the ambition and determination of their original curators as well as the contribution of many donors. Morgan contends that embryo collections were instrumental in shaping what she calls an "embryological vision of the world", a modern vision of human pregnancy as a series of developmental stages, revealed by technological means. She argues that embryology has been so influential, and the embryological vision produced so taken for granted, that images of the embryo and foetus are frequently used to stand in for "life" itself. Embryo collections, Morgan argues, enabled the classification of pregnancy from the "point of view" of the developing embryo and foetus, and their legacy today is in part revealed by the affective power of images of the foetus in public culture.[5] This embryological vision of the world is critiqued by feminist scholars in the U.S. and Europe for its promotion of the foetus as "free-floating" and independent of the mother's body (Petchetsky 1987, Roberts 2012, see also Buklijas and Hopwood 2008).

Although the collection of embryos to illustrate human development no longer takes place at the scale described by Morgan and made manifest in the Kyoto Collection, embryos and the stem cells derived from them continue to be viewed as controversial biological materials engendering public debate and regulatory oversight. Morgan cites the iconic status of the foetus as a figure of "life", beginning with the presentation of foetal imagery in the documentary photography of Lennart Nilsson (1965) that featured on the cover of *LIFE* magazine. Pro-life campaigns use images of embryos and foetuses to argue for the restriction or prohibition of abortion; foetal and embryonic imagery features heavily in medical representations of pregnancy as well as pregnancy advice literature; and "fetishized fetal imagery" is used to market an array of products to consumers (Paxson 2004, 246, see also Petchetsky 1987, Haraway 1997, Morgan and Michaels 1999, Taylor 2008). The ubiquity of foetal imagery in American popular culture and in the debates surrounding women's access to abortion suggests that the remaining collections need to be maintained. These collections offer historical insight into the development of a science and the knowledge practises that relied so significantly on the contributions, even unwitting, from the pregnancies of thousands of women.

Embryo collections also played an important role in grounding theories of human development, and in materially underpinning the study of groups of people as "populations". The ethical and relational dimensions of collecting bodily materials are the subject of extensive work in sociological and anthropological literatures. However, the spatial dimensions of the establishment of tissue collections as exemplars of "populations", defined in ethno-racial terms, or as representatives of human population diversity, have only recently begun to receive attention from geographers (see Nash 2015). Despite their role in shaping medical and scientific knowledge, embryological, anatomical and other collections of human biological materials have been of relatively little interest to cultural or medical geographers to date, with a few notable exceptions (Hill 2006, Hussey 2017, Morton 2017). Indeed, there is far more detailed research in historical and cultural geography on natural history collections focusing on the lives of animals and plants, and the specimens they became, than on medical and scientific exploration of the human body's interior. Yet these interior bodily spaces, and the efforts to make them visible in anatomical collections, are critical to scientific as well as popular understandings of the relationship between life and death, the individual and the population, and in the embryo collection, the pregnant body and the foetus.

Studies of the creation of human anatomical collections, including embryo collections, can reveal how environmental and behavioural dimensions of human existence, and the relations between them, were theorised and put into scientific practice. More importantly, anatomical collections and particularly collections that seek to archive human development through materials derived from pregnancy suggest that the collection, classification and analysis of biological materials from the body's interior were also central spatial concerns of modern biopolitics. Indeed, the study of life in the medical and life sciences was never confined to that of the living organism but also to the fixing or arresting of life, preserving and retaining dead specimens for future study in the spaces of scientific collections, laboratories and biobanks (Radin 2013, Fannin and Kent 2015). Tissue collections were one of the means by which reproductive bodies could reveal their innermost secrets. Connecting the interior of the pregnant body to the study of a national "population", they also helped create norms through which the surveillance of pregnancy could operate.

Notes

1 In the United States, the Carnegie Institution of Washington houses over 8,000 embryonic and fetal specimens collected from 1911 to the 1940s. The U.S. National Museum of Health & Medicine's Human Developmental Anatomy Center also houses several smaller collections. In Europe, the University of Göttingen is home to one of the most striking collections, the Blechschmidt Collection established in 1948, that includes large-scale reproductions in wax of human embryos at various stages of development. Other significant embryo collections in Europe include the Hinrichsen (Ruhr-University Bochum), the Madrid Collection (Complutense University of Madrid), the Hamilton-Boyd Collection (Cambridge University), the Hubrecht Collection (Museum für Naturkunde, Berlin) and the Human Developmental Biology Resource (Newcastle University).

2 Nishimura published widely in English-language journals, although a review of his 1964 publication entitled *Chemistry and Prevention of Congenital Anomalies* (Springfield, IL: Charles C. Thomas) lamented that the text was "devoted almost exclusively to American and European work" and offered few insights into the "extensive work which must be presumed to be going on in a country which might well claim supremacy in biological laboratory work, facilities and equipment, and in which abortion is legal." See J. H. Edwards, *Proceedings of the Royal Society of Medicine*, 1964 November, 57(11): 1118.

3 The history of Japanese laws related to abortion and contraception note the exceptional status of abortion law in the post-war period compared to other places. Access to abortion in the U.S. and Europe during this same period was very difficult to obtain and in the U.S. was particularly limited. By 1967, 49 U.S. states and Washington, D.C. regarded abortion as a felony and access to information about abortion was severely restricted under the Comstock Act of 1873. The liberalisation

of abortion in both the U.S. and across much of Europe did not take place until the 1960s and 1970s. The Comstock Act was repealed in 1971, the same year in which abortion in the U.S. was "legalised" although not granted as an absolute right to women. Rather, access to abortion in the U.S. under Roe v. Wade was granted on the basis of protecting rights of privacy in the doctor–patient relationship.

4 This invisibility of women's embryo donation is complicated in the case of donation originating from the termination of pregnancy, as there is the presumption that women do not want to know about the use of their donated tissues (Kent 2008). Women do express ambivalence about the use of their tissues: some are keen to know that the material they donate will not continue to "live" and are therefore uncomfortable with use of this material in research. Others by contrast are interested in seeing their material "live on" as useful in another (research) setting. Kato's study of women's embryo donation during IVF reflects a similar ambivalence on the part of donors as well as changing conceptions of the embryo as women went through the process of IVF. Some wanted to "protect" their embryos from use by researchers; others expressed their willingness to donate as a way to "give meaning to the lives of their 'children,' or embryos, just as some parents might donate organs of their deceased children so that they might contribute to other people's lives" (Kato 2014, 361).

5 The post-war imaginary of the "menacing" foetus that haunts its mother captivated audiences and consumers of new spiritual practices of *mizuko kuyo* in Japan in the 1970s. The cultural and religious dimensions of these practices carried out to appease the spirits of deceased foetuses are explored in works by Helen Hardacre (1997) and William LaFleur (1994).

References

Asai, Atsushi, Motoki Ohnishi, Etsuyo Nishigaki, Miho Sekimoto, Shunichi Fuku-hara, and Tsuguya Fukui. 2002. "Attitudes of the Japanese Public and Doctors towards Use of Archived Information and Samples without Informed Consent: Preliminary Findings Based on Focus Group Interviews." *BMC Medical Ethics* 3 (1). http://doi:10.1186/1472-6939-3-1.

Bashford, Alison. 2007. "Nation, Empire, Globe: The Spaces of Population Debate in the Interwar Years." *Comparative Studies in Society and History* 49(1): 170–201.

Buklijas, Tatjana, and Nick Hopwood. 2008. *Making Visible Embryos*, www.hps.cam.ac.uk/visibleembryos, accessed 18 June 2018.

Coleman, Samuel. 1992. *Family Planning in Japanese Society: Traditional Birth Control in a Modern Urban Culture.* Princeton: Princeton University Press.

Congenital Anomaly Research Center. 2010. Kyoto Human Embryo Visualisation Project. Kyoto University Graduate School of Medicine. http://bird.cac.med.kyoto-u.ac.jp/index_e.html, accessed 1 August 2017.

Dickenson, Donna. 2007. *Property in the Body: Feminist Perspectives.* Cambridge: Cambridge University Press.

Fannin, Maria, and Julie Kent. 2015. "Origin Stories from a Regional Placenta Tissue Collection." *New Genetics and Society* 34(1): 25–51.

Fujimoto, Toyoaki. 2001. "Nishimura's Collection of Human Embryos and Related Publications." *Congenital Anomalies* 41: 67–71. doi:10.1111/j.1741-4520.2001.tb00875.x

Haraway, Donna J. 1997. "The Virtual Speculum in the New World Order." *Feminist Review* 55(1): 22–72.

Hardacre, Helen. 1997. *Marketing the Menacing Fetus in Japan.* Berkeley, CA: University of California Press.

Hill, Jude. 2006. "Travelling Objects: The Wellcome Collection in Los Angeles, London and Beyond" *cultural geographies* 13(3): 340–366.

Hill, Mark, Kohei Shiota, Shigehito Yamada and Cecilia Lo. 2016. *Kyoto Embryo Collection*. UNSW Australia: iBooks. https://itun.es/us/faUleb.l

Homei, Aya. 2012. "Midwives and the Medical Marketplace in Modern Japan" *Japanese Studies*. 32(2): 275–293.

Hussey, Kristin D. 2017. "Seen and Unseen: The Representation of Visible and Hidden Disease in the Waxworks of Joseph Towne at the Gorden Museum." *Interdisciplinary Studies in the Long Nineteenth Century*. 19(24) http://doi.org/10.16995/ntn.787

Kameda, Tomomi, Shigehito Yamada, Chigako Uwabe, and Nobuhiko Suganuma. 2012. "Digitization of Clinical and Epidemiological Data from the Kyoto Collection of Human Embryos: Maternal Risk Factors and Embryonic Malformations." *Congenital Anomalies* 52(1): 48–54.

Kanahashi, Tohoru, Shigehito Yamada, Mire Tanaka, Ayumi Hirose, Chigako Uwabe, Katsumi Kose, Akio Yoneyama, Tohoru Takeda, and Tetsuya Takakuwa. 2016. "A Novel Strategy to Reveal the Latent Abnormalities in Human Embryonic Stages from a Large Embryo Collection." *The Anatomical Record* 299(1): 8–24.

Kato, Masae and Margaret Sleeboom-Faulkner. 2011. "Meanings of the Embryo in Japan: Narratives of IVF Experience and Embryo Ownership." *Sociology of Health and Illness* 33(3): 434–447.

Kato, Masae. 2014. "Giving a Gift to the Gift: Women's Experiences of Embryo Donation in Japan." *Anthropological Forum* 24(4): 351–363.

Kent, Julie. 2008. "The Fetal Tissue Economy: From the Abortion Clinic to the Stem Cell Laboratory." *Social Science & Medicine* 67(11): 1747–1756.

LaFleur, William R. 1994. *Liquid Life: Abortion and Buddhism in Japan*. Princeton: Princeton University Press.

Lenz, W. 1988. "A Short History of Thalidomide Embryopathy" *Teratology* 38: 203–215.

Levine, Philippa and Alison Bashford. 2010. "Introduction : Eugenics and the Modern World" in *The Oxford Handbook of the History of Eugenics*, edited by Alison Bashford and Philippa Levine, 3–24. Oxford: Oxford University Press

Morgan, Lynn M. 2004. "A Social Biography of Carnegie Embryo No. 836." *The Anatomical Record (Part B: The New Anatomist)* 276B: 3–7.

Morgan, Lynn. 2009. *Icons of Life*. Berkeley: University of California Press.

Morgan, Lynn M. and Meredith W. Michaels, eds. 1999. *Fetal Subjects, Feminist Positions*. Philadelphia: University of Pennsylvania Press.

Morton, Sarah. 2017. The Legacies of the Repatriation of Human Remains from The Royal College of Surgeons of England. PhD diss., University of Oxford.

Nagai, Momoko, Katsura Minegishi, Munekazu Komada, Maiko Tsuchiya, Tomomi Kameda, and Shigehito Yamada. 2016. "Extraction of DNA from Human Embryos after Long-term Preservation in Formalin and Bouin's Solutions." *Congenital Anomalies* 56(3): 112–118.

Nash, C. (2015) *Genetic Geographies: The Trouble with Ancestry*. Minneapolis, MN: Minnesota University Press.

Nilsson, Lennart. 1965. "Drama of Life Before Birth." *Life*, April 30.

Nishimura, Hideo, Kiichi Takano, Takashi Tanimura and Mineo Yasuda. 1968. "Normal and Abnormal Development of Human Embryos: First Report of the Analysis of 1,213 Intact Embryos." *Teratology* 1: 281–290.

Nishimura, Sey. 1996. "Hideo Nishimura, MD." *JAMA* 275(14): 1137. doi:10.1001/jama.1996.03530380079044

Norgren, Tiana. 2001. *Abortion before Birth Control: The Politics of Reproduction in Postwar Japan*. Princeton and Oxford: Princeton University Press.

Ogino, Miho. 1996. "Abortion, the Eugenic Protection Law, and Women's Reproductive Rights in Japan. *Atlantis* 21(1): 133–138.

Otsubo, Sumiko. 2005. "The Female Body and Eugenic Thought in Meiji Japan" in *Building a Modern Japan: Science, Technology and Medicine in the Meiji Era and Beyond*, edited by Morris Low, 61–81. New York: Palgrave.

Parry, Sarah. 2006. "(Re)constructing Embryos in Stem Cell Research: Exploring the Meaning of Embryos for People Involved in Fertility Treatments", *Social Science & Medicine* 62(10): 2349–2359.

Paxson, Heather. 2004. *Making Modern Mothers: Ethics and Family Planning in Urban Greece*. Berkeley: University of California Press.

Petchetsky, Rosalind Pollack. 1987. "Fetal Images: The Power of Visual Culture in the Politics of Reproduction." *Feminist Studies* 13(2): 263–292. doi:10.2307/3177802.

Radin, Joanna. 2013. "Latent Life: Concepts and Practices of Human Tissue Preservation in the International Biological Program." *Social Studies of Science* 43 (4): 484–508.

Reubi, David. 2018. "A Genealogy of Epidemiological Reason: Saving Lives, Social Surveys and Global Population." *BioSocieties* 13(1): 81–102.

Roberts, Julie. 2012. *The Visualised Foetus: A Cultural and Political Analysis of Ultrasound Imagery*. Farnham: Ashgate.

Robertson, Jennifer. 2010. "Eugenics in Japan: Sanguinous Repair" in *The Oxford Handbook of the History of Eugenics* edited by Alison Bashford and Philippa Levine, 430–448. Oxford: Oxford University Press.

Suzuki, Michiko. 2013. "Fat, Disease and Health: Female Body and Nation in Okamoto Kanoko's 'Nikutai no shinkyoku'" *U.S.-Japan Women's Journal* 45: 33–49.

Tanimura, Takashi. (1996) "In Memoriam Dr. Hideo Nishimura (1912–1995)." *Congenital Anomalies* 36: 2–5. doi:10.1111/j.1741-4520.1996.tb00315.x

Taylor, Janelle S. 2008. *The Public Life of the Fetal Sonogram: Technology, Consumption, and the Politics of Reproduction*. New Brunswick, NJ: Rutgers University Press.

Thompson, Charis. 2007. *Making Parents: The Ontological Choreography of Reproductive Technologies*. Cambridge, MA: The MIT Press.

Tsuge, A. 2002. "Sentaigijutsu ga 'juyô' sareru toki: ES saibôkenkyû no jirei kara [When Advanced Technology is Accepted: Cases from hESC Research]" *Gendai shisô* 30(2): 76–89.

Waldby, Catherine and Robert Mitchell. 2006. *Tissue Economies: Blood, Organs, and Cell Lines in Late Capitalism*. Durham, NC: Duke University Press.

Yamada, Shigehito, Chigako Uwabe, Tomoko Nakatsu-Komatsu, Yutaka Minekura, Masaji Iwakura, Tamaki Motoki, Kazuhiko Nishimiya, Masaaki Iiyama, Koh Kakusho, Michihiko Minoh, Shinobu Mizuta, Tetsuya Matsuda, Yoshimasa Matsuda, Tomoyuki Haishi, Katsumi Kose, Shingho Fujii, and Kohei Shiota. 2006. "Graphic and Movie Illustrations of Human Prenatal Development and Their Application to Embryological Education Based on the Human Embryo Specimens in the Kyoto Collection." *Developmental Dynamics* 235(2): 468–477.

2 Biological reproduction, respatialised

Conceiving abnormality in a biotech age

Juliane Collard

Introduction

Since its clinical debut in humans in 1978, *in vitro* fertilisation (IVF) has become a routine medical procedure. Alongside IVF, a broad suite of assisted reproductive technologies (ARTs) has shifted from the theoretical to the commonplace. Loosely defined as technological tools used to manipulate human gametes or embryos for the purposes of establishing a pregnancy, ARTs belong to a class of biotechnologies directed at capturing the reproductive potential of cells by causing them to live differently in space and time (Landecker 2005). Once confined to the womb, fertilisation and early embryo development now occur readily and with increasingly high success rates *in vitro*, opening up the biology of fertility to flexible spatial and temporal possibilities. As Hannah Landecker (2006, 32) writes, these technologies have "enabled life to be extracted from the body" where it can survive, regenerate, move and grow. *Ex vivo* gametes and embryos can be manipulated and biopsied; genetically screened and diagnosed; frozen, stored for long periods and thawed again for implantation in the womb; donated to research or to another infertile couple; immortalised as cell lines and disease models and circulated as mobile proxies for the bodies from which they derive (Parry 2012). As I unpack in greater detail later in this chapter, these biotechnological interventions have facilitated significant advances in biomedical research, for example, through the use of human embryonic cells as surrogates for whole human bodies in genetic testing and disease modelling.

Geographers have recently become interested in some of these spatial dynamics, paying especial attention to transnational surrogacy and cross-border markets in reproductive care and politics (Parry 2012; Greenhough et al. 2015; Payne 2015; Lewis 2016, 2017; Schurr 2016; Schurr and Militz 2018). Work in this vein has primarily focused on the *extensification* of reproductive relations and economies across space and time. New biotechnological developments have facilitated the emergence of a global trade in bodily commodities. Eggs, sperm and embryos travel between bodies, across borders and around the world faster and more cheaply than ever before. Researchers have attended closely to the circulation of power in these extensive reproductive networks, examining the processes through which some bodies and lives are rendered more bioavailable (Cohen 2005)[1] as surrogates and tissue "donors" along well-worn lines of, *inter alia*, race, class, gender and geography.

Less attention has been paid, however, to the geographical *intensification* facilitated by ARTs. The respatialisation of reproduction from within to outside the body has made reproduction hypermobile, but it has also rendered reproductive tissues hyper-visible at the most intimate scales of resolution. In our current biotech age, biological life is studied, mapped and manipulated in previously impossible ways. Assisted reproductive technologies grant clinicians, scientists and intended parents unprecedented, if always incomplete, access to the concrete space of the embryo itself. Products with names like GenePeeks, InSight and RiWitness allow embryologists and fertility specialists to make the previously invisible visible, to examine genes and chromosomes at the earliest stages of human life. In a beautiful scalar metaphor, one clinician I spoke with compared new ARTs to the Hubble telescope, suggesting that they allow us to "see things never before seen with the human eye".

Drawing on six months of fieldwork I conducted in the U.S. between 2016 and 2017, this chapter follows the clinical gaze into the genetic and chromosomal recesses of the human IVF embryo. Taking my cue from the embryologists, geneticists, fertility doctors, stem cell researchers, life scientists, bioethicists and lawyers with whom I conducted interviews, and from my observations at major scientific expos and reproductive medicine conferences, I orient myself toward abnormality in this space. The detection and diagnosis of embryonic abnormality is increasingly central to contemporary fertility science and practice. The biotech products mentioned above are explicitly geared toward the identification of abnormal embryos so that they can be eliminated from the reproductive stream. Conversations and research about abnormal embryos and their preimplantation detection predominate in the fertility clinic, at conferences on fertility and assisted reproduction and in bioethical debates over the use and regulation of assisted reproduction technologies.

In a compelling geographical twist, the detection of embryonic abnormality *in vitro* relies on the use of detailed biological maps—visual representations of the absolute and relative places of genetic sequences on the 23 pairs of chromosomes in each human cell. Clinicians depend on these maps as locational tools, means of navigating the biological terrain of life itself. The "geography of genes" (Hall 2003, 152), it would seem, is essential to grasping the gene's role in human development, to understanding how we are assembled. By connecting bodily (mal)function to particular stretches of the genetic sequence, gene maps ostensibly make it possible to predict future health outcomes, even at the earliest stages of life. As James Watson, one of the co-discoverers of DNA, said almost three decades ago: "We used to think our fate was in our stars. Now we know, in large measure, our fate is in our genes" (quoted in Jaroff 1989, 67). This idea of the body as readable, mappable terrain is by no means new (Foucault 1989;

Waldby 2000). Its application in assisted reproduction, however, has the potential to radically and unevenly alter our socio-biological futures.

I proceed from here in four parts. Each is an attempt to place the abnormal embryo in the dense web of integrations (Haraway 1997) within which it is produced and to flesh out the work required to sustain it. The abnormal embryo is, I suggest, as much a social as a biological entity—the product of a host of technical, legal, scientific, affective and political practices. The first section briefly overviews the respatialisation of reproduction with which I opened: from *in* to *ex vivo*. In the second section, I turn to the detection of embryonic abnormality *in vitro*, and to the importance of gene maps in genetic screening. I focus on the use of preimplantation genetic tests (PGTs)—reproductive technologies used to screen IVF embryos for genetic and chromosomal abnormalities before they are transferred into the womb. By locating and localising abnormality within the embryo, I suggest, PGTs render abnormality as an internal biological feature of life. The third section queries how the knowledge generated by PGTs is instrumentalised—used to understand and order human life. The fertility clinic, I suggest, becomes a critical site in the negotiation and materialisation of the border between the normal and the abnormal, with explicitly material implications. In the fourth and final section I follow abnormal embryos as they are surplussed out of the fertility clinic and into the research lab, where they comprise a unique source of biovalue.

From *in vivo* to *in vitro*: The respatialisation of reproduction

Developed first in invertebrates, then amphibians, reptiles and fish and eventually in non-human mammals, the turn to human clinical IVF was neither straightforward nor simple. Like many of the biotechnological innovations that have preceded it, it was built with tools developed over centuries, by generations of scientists and researchers, across a wide range of disciplines, with multiple and disparate objectives (Franklin 2013). More than a few of the major techniques used during IVF were discovered by accident, including the ability to freeze mammalian gametes (Gordon 2003), a precondition to the preimplantation genetic tests that lie at the heart of this chapter. It wasn't until 1978 that IVF was successfully translated into human clinical application. That year, the world was introduced to *ex vivo* sexual reproduction with the birth of Louise Brown, the first "IVF baby". Its rapid expansion in the 1980s and 1990s produced a new generation of so-called "miracle babies" born to infertile couples in countries throughout the Western world.

In addition to establishing a new method of reproduction, IVF has provided researchers and clinicians with a powerful window onto early human development. Since the 1980s, clinical IVF has generated a reliable supply of *in vitro* research embryos. Although *ex vivo* embryo research has a much longer history, dating back to the late nineteenth century (Morgan 2009), today's research embryos differ in an important way: they retain their vital, regenerative, reproductive capacities (Landecker 2005; 2006). In concert with new methods of keeping embryos alive outside the body, such as tissue culturing and cryopreservation, IVF has allowed scientists to explore the basic mechanisms of human conception, heredity and development. By harnessing embryos' productive and reproductive potential, researchers can study foundational cellular processes in real-time, during crucial stages of embryo development (Franklin 2013). As one researcher told me, "In human systems you can't get access to early embryos. So there's this big gap in knowledge. We address that by using early embryos themselves and studying them *in vitro*, you know, in a dish, and using those as models for growth and development.... You can take an early embryo, put it under a microscope, pluck out one cell, and just study it" (personal communication).

No longer hidden in the fleshy, opaque depths of the whole female womb, lively *ex vivo* embryos can be dissected and diagnosed, turned into self-replicating cell lines, and used as disease-in-a-dish models for the study of pathogenesis. Forty years after Louise Brown's birth, the respatialisation of reproduction preconditioned by IVF has provided the stage for myriad feats in basic science and reproductive biology. Onco-mice, Dolly the sheep, induced pluripotent stem cells and a menagerie of other admixed chimeras, clones and cellular tools owe their existence to *ex vivo, in vitro* fertilisation (Franklin 2013).

Of particular importance to this chapter, IVF has provided a platform for preimplantation genetic tests, and thus for the detection of abnormality in early embryos. As discussed in more detail below, PGT techniques require that a sample of cells be "plucked" from an *in vitro* embryo—cells which are tested for genetic and chromosomal abnormalities before the embryo is implanted in the womb. It is into these cells that the clinical gaze peers, looking for signs of abnormality in the genes and chromosomes it finds there.

Mapping embryonic abnormality

Preimplantation genetic tests comprise one genre of assisted reproductive technologies. Primarily diagnostic tools, they use biopsy techniques to remove a sample of cells from a five- to six-day old embryo. The technique relies on what Siddhartha Mukherjee (2016, 456) calls "a peculiar idiosyncrasy of human embryology". If a sample of cells is removed from an embryo at just the right time, the remaining cells divide to fill the gap. "For a moment in our history", Mukherjee (2016, 456) writes, "we are actually quite like salamanders, or, rather, like salamanders' tails—capable of complete regeneration even after being cut by a fourth". The human embryo can thus be biopsied at an early stage, a few cells extracted for use in genetic tests, without compromising the embryo's capacity to develop into human life.

Following biopsy, the "cut" embryo is cryopreserved. The cells removed are tested for chromosomal or genetic abnormalities, becoming proxies for the whole embryo from which they were derived. Although the validity of these tests is contested in both scientific literature and popular media, they provide the basis for decisions about embryo categorisation, as normal or abnormal, and disposition, whether the embryo will be implanted or discarded. As discussed in the next section, the majority of clinics will not proceed with the transfer of an abnormal embryo.

There are two basic types of PGTs: preimplantation genetic diagnosis (PGD) and preimplantation genetic screening (PGS). Both rely on biological maps as a guide to the genomic landscape of the embryos being tested. The relative and absolute locations of genes and chromosomes and the clinical significance of these locations—knowledge generated through gene mapping—precondition both the prediction of biological abnormalities and their effect on the body. Fertility clinics thus use these maps as a means of directing and interpreting screening results and of selecting the "healthiest" embryo for implantation.

Preimplantation genetic diagnosis is used to search for and locate specific genetic traits inside the embryo by examining a sample of its cells. Intended parents using PGD may have experienced recurrent pregnancy loss. They may want to select an embryo with a particular blood or bone marrow type to create a future donor, or "saviour sibling", for an existing ill child. They may want to choose an embryo carrying a genetic trait that they themselves carry, for example deafness or dwarfism.[2] Or they may have been identified as high-risk for a single gene disorder based on personal or familial factors. "They may have BRCA[3] in the family, or Cystic Fibrosis, or Huntington's Disease. They access assisted reproduction to take that out of the family line," one interviewee explained.

Presently, labs can screen for upwards of 1,000 single-gene disorders and risk factors, a number that is rapidly increasing. Disorders are added to the list as their absolute location within the genomic landscape is determined, a mapping endeavour that requires significant financial and scientific inputs. Without adequate geographical information, however, PGD would be futile. Put simply, gene maps tell clinicians where to look: for Huntington's in a particular stretch of chromosome four, for Tay-Sachs on chromosome 15 and so on. As the number of mapped genes increases, so too do the clinical applications for PGD: once a gene has been located in the genetic landscape it becomes possible to screen for it.

Although relatively rare, the use of PGD is on the rise as fertility clinics begin to include carrier genetic testing[4] and multi-generation family medical histories within their remit. Under the logic of PGD, carriers of heritable genetic conditions are effectively infertile—incapable as they are of reliably reproducing genetically normal, "healthy" babies—and thus require medical intervention. And in reproductive medicine, everyone is considered to be a potential carrier. "Every person has an average of six recessive mutations that could cause genetic diseases in future generations", a GenePeeks brochure informs its readers.

In response to this risk, some clinics have made personalised genetic screening a precondition to fertility treatment. "Everybody speaks to a genetic risk assessor", the director at a Bay Area clinic told me. "It's mandatory. And we do the 24-panel [genetic test] as well as looking at a four generation pedigree [family history]. We've picked up on all sorts of things.... Any disease that comes along, if we can sequence the gene, we can test the embryo for it". The result of these extensive personalised screening practices is a patient population carrying relatively high rates of detected single-gene disorders and genetic risk factors, both of which trigger the use of PGD. By sequencing the intended parents' genome, clinicians and embryologists equip themselves with a spatialised genetic knowledge that allows them to focus their attention on a particular location of the embryo's genetic code.

If PGD is a diagnostic tool, preimplantation genetic screening (PGS) is a means of prioritising embryos for transfer. As described to me by the practitioners who use them, patients access PGS to select the embryo with "the highest chance of getting them a healthy baby" (fertility specialist, Bay Area). Used to comprehensively screen the chromosomal makeup of a given embryo, PGS comprises the vast majority of preimplantation genetic tests conducted. Whereas PGD relies on maps that depict the absolute location of genes in the genetic code, PGS actively detects and maps whole chromosomal aberrations. Test results are presented visually: graphs that chart the presence of chromosomal abnormalities and where they occur. Again, disease is linked to a precise, biological location in the cell: to an extra copy of chromosome 21 (Down Syndrome) or to a single copy of the X chromosome (Turner Syndrome), for example. The biological maps produced during PGS become physical, visual evidence of anomalous development, confirmation that our fate is, indeed, in our genes.

In some clinics, the use of PGS is restricted to patients of high maternal age or with other risk factors. In others, its use has been routinised, a hallmark feature of what Charis Thompson (2013) has called today's selecting society. Many clinics, for example, offer it as the most effective means of sex-selection, or "family balancing" as it is more euphemistically called in the industry. In California, a world leader in assisted reproduction and the state with which I became most familiar during my fieldwork, the average screening rate is around 60 per cent, meaning that 60 per cent of IVF embryos are screened using PGS. In some clinics, notably those that have mandatory screening practices, this number can skyrocket to over 95 per cent. As the technology becomes more "widely available, easier to access and more cost effective, people are developing a comfort with it", one clinician told me.

In both preimplantation genetic diagnosis and preimplantation genetic screening, biological maps perform crucial work. By linking physiological conditions to precise biological locations, they effect a secondary spatialisation of pathological development akin to that discussed by Foucault in *The Birth of the Clinic* (1989). Under the rubric of gene mapping, abnormality finds its seat in corporeal space, "becoming visible in a geographical system of masses" (Foucault 1989, 9). These maps thus become central to the reproduction of abnormality as a biological fact of life—something that is "there" to be discovered. By directing the clinical gaze to a particular stretch of genetic code or to a specific chromosome, they suggest the possibility of predicting the biological future. Although different in format from cartographic representations of land and water, gene maps likewise imply the possibility of mastery over nature (Wald 2000; Zwart 2009), rendering our internal biology as legible, ordered territory. Abnormality is naturalised and materialised as a feature of our genomic landscape. Its broader social, political and economic dimensions fall away.

As gene-sequencing technologies become cheaper, faster and easier to use, the hunt for particular genes—for intelligence, sociability, infidelity, sexuality, height, life expectancy and so on—has intensified dramatically. Emerging biotech firms jostle for position by promising the genomic coordinates for characteristics like IQ and athleticism—projects that see start-ups mapping the whole genomes of mathematical geniuses (Regalado 2017) and star athletes (Kayayerli 2017) in an attempt to find the precise location of their genetic superiority. As more and more genes are mapped, a host of physiological and psychological conditions becomes visible, read as a set of "displaced, destroyed, or modified elements bound together in a sequence according to a geography that can be followed step by step" (Foucault 1989, 136; see also Philo 2000). On the basis of this knowledge, biotech firms advertise plans to offer disease risk reports on every IVF embryo, a bid to aid parents and clinicians in reproducing the best possible babies (Regalado 2017). Moving well beyond the detection of fatal and severe genetic abnormalities, a whole array of human characteristics stands to be biologised and pathologised. Once discovered, these genes may be targeted for (de)selection through preimplantation screening.

Although many of the selection technologies peddled by the biotech sector remain anticipatory, imagined next steps to PGTs, they have material effects in the present. Huge financial resources are invested in research and development. Fertility doctors field questions from intended parents about selecting embryos based on a variety of conditions. Intended parents themselves increasingly undergo genetic screening prior to attempting to conceive. And the field of assisted reproduction remains steadfastly fixed on the genetic and biological dimensions of human health. Meanwhile, the number of PGTs conducted continues to rise.

Negotiating abnormality in the fertility clinic

Each time a preimplantation genetic test is run, two related determinations are made: first, whether the embryo is normal or abnormal, and second, whether the embryo should be implanted or discarded. These tests are, in effect, powerful boundary making tools—a means of drawing a line between the normal and the abnormal as distinct, natural kinds. Seated within the embryo's genes and chromosomes, abnormality is operationalised as a natural fact: an *a priori*, biological status.

But as Foucault (2003) has so productively explored in his lectures on abnormality, as elsewhere, both the normal and the abnormal are inherently political concepts, elements on the basis of which the exercise of power is founded and legitimated. The construction of a category of abnormal embryos in the fertility clinic evinces such an exercise of power. The deployment of abnormality as a naturalised interior quality of the embryo camouflages the power relations that materialised them as natural and abnormal in the first instance. As I discuss further below, decisions about which abnormalities are targeted for eradication, which embryos are cast aside, belie deeply held assumptions about the bodies we are supposed (to want) to have and reproduce. Put otherwise, those putatively natural abnormalities targeted for de-selection are the product of (among other things) disciplinary knowledge/power, affect, scientific, medical and administrative practices, legal regimes and a host of other material and discursive relations (Tremain 2006).

Among other outcomes, including an advertised (though contested) increase in normal live births, the use of PGS and PGD has resulted in a high volume of embryos being diverted away from the reproductive pathway: embryos with single gene disorders, genetic risk factors, or chromosomal abnormalities; those which are ambiguously sexed or of the undesired sex and, less commonly, those with the wrong blood or bone marrow type who will be incompatible as saviour siblings. Embryos with each of these characteristics fall on the wrong side of the biopolitical border established through preimplantation genetic testing—not normal, they can only be abnormal.

In some cases, such as during sex selection, this diversion may be temporary. The undesired embryo might be cryopreserved for future use by the intended parents or, less commonly, donated to another infertile couple (see Cromer 2017).[5] The genetically or chromosomally abnormal or mosaic embryo, however, is almost always cast permanently out of the reproductive stream. Most clinics have strict policies in place to prohibit the transfer of abnormal embryos, rendering them unavailable for reproductive use. Abnormality is taken to interrupt what Catherine Waldby and Susan Squier (2003) call biographical embryology: a developmental narrative in which the human body is imagined to emerge ineluctably from the embryological process.[6]

Importantly, an embryo's viability, its "compatibility with life" to use the language of my interlocutors, is almost immaterial here. In a collapse of the abnormal with the non-viable, embryos that screen positive for non-fatal conditions[7] are subject to the same prohibitive policies, sometimes regardless of the intended parents' wishes.[8] "We will not take that liability risk of implanting [abnormal embryos] and then somebody coming back to us and saying, you know, 'this baby isn't normal'. We just don't take any risk with it," said one embryologist. In the fertility clinic, the norm becomes the rule (Canguilhem 1991).

This collapse suggests an understanding of abnormality that extends beyond its biological meaning. Decisions about which embryos to implant and which to avert are steeped in socio-cultural ideas of normativity, able-bodiedness, health, sexual dimorphism and neuro-typicality: in short, of idealised form and function. Practitioners debate, for example, whether or not parents should be allowed to transfer female embryos with the gene predisposing them to breast cancer (BRCA), or male embryos with an extra x chromosome. In each of these decisions, the border between the normal and the abnormal is negotiated, drawn and reproduced. On the basis of these categories, some embryos are implanted while others are discarded. These categories thus become sites for the exercise of power that result not only in the distribution of resources, but also of life itself. In the fertility clinic, the abnormal embryo has no biographical future: it cannot be let to develop into a full human life.

The prohibition against transferring abnormal embryos evinces an assumed-to-be-shared social opinion about the non-viability and non-desirability of particular kinds of biological difference. Discourses and practices of reproduction are shot through with profound anxieties about the risk of abnormal foetal development (Oakley 1984; Lippman 1991; Duden 1993; Kukla 2005; Tremain 2006; Mansfield 2012; Kafer 2013). These anxieties materialise in the regularisation of PGT, in the use-histories of more common screening technologies like ultrasound (Duden 1993) and amniocentesis (Rapp 1999), and in the development of reproductive technologies themselves. It would be a "sin of parents to have a child that carries the heavy burden of genetic disease," said Robert Edwards, awarded the Nobel Prize in medicine in 2010 for his pioneering work on IVF. "We are entering a world where we have to consider the quality of our children," he went on.[9] Writes feminist crip scholar Alison Kafer (2013, 29), "reproductive futurity demands a Child that both resembles the parents and exceeds them; 'we' all want 'our' children to be more healthy, active, stronger, and smarter than we are, and we are supposed to do everything in our power to make it happen". To want a disabled child, to even accept disability, is seen as disordered and unbalanced (ibid.).

Although understood in the clinic as a way to improve intended parents' chances of carrying a healthy baby to term, disciplinary embryo transfer practices echo with historical and contemporary oppressions. Assisted reproduction has a messy genealogy that ties it, chronologically and biopolitically, to periods of intense discrimination. Whole social groups—categorised as "mentally ill", "impaired", "defective", or "degenerate", among other labels used to demarcate abnormality—have faced sterilisation, segregation, incarceration and institutionalisation; violence and abuse and the withholding of the rights of citizenship (Ordover 2003; McWhorter 2009; Kafer 2013). As a clinical practice and a scientific project, preimplantation genetic testing makes troubling claims of differential life worth on the basis of biological differences. The proliferation of these tests, then, might be seen as consolidating and reproducing what Kafer (2013), following McRuer (2006), calls the compulsory reproduction of able-bodiedness and able-mindedness. The field of appropriate conduct in response to pregnancy is narrowed. Intended parents, and women in particular, are urged or even required to take advantage of screening technologies and are thus enlisted to facilitate the normalisation of the embryo, the foetus, the child.

The socio-biological boundary-making practices under discussion here have produced a large store of embryos seen to have no future life potential. Although of little value to the fertility clinic, these embryos are readily constructed as an invaluable *scientific* surplus whose regenerative capacities should not be wasted (Parry 2006; Waldby and Cooper 2010). Donation to biomedical research is framed as a way to transform this potential waste into new kinds of value. "What we have been finding in patients with excess embryos," a stem cell researcher told me, "is that they want to *do* something *useful* with them. They want to give them to research. So I get on my phone, right here, at least once a month, either a patient or a physician calling me to say 'I know you have a stem cell program. I have a patient. They want to give their embryos to research'" [emphasis his]. Sarah Franklin and her colleagues noted a similar narrative at work in the U.K. Their study at a PGD centre found that 67 per cent of patients were willing to donate embryos to research, of whom more than 80 per cent expressed a desire to "give something back" (Franklin et al. 2005; see also Franklin 2013).

In this context, abnormal embryos form a reserve of highly valuable biological research material. Although data on disposition decisions is limited—clinics in this study did not gather statistics regarding the disposition of normal versus abnormal embryos—anecdotal evidence suggests that abnormal embryos are more likely to be donated to biomedical research than their chromosomally and genetically normal counterparts (see also Franklin 2013). Because the embryo market remains effectively illegal in the U.S.—unlike oocytes, embryos cannot be bought and sold—these donated embryos are essential to continuing biomedical research. Although donors cannot claim property rights in their own tissues, once "gifted" by the donors to the research lab, they become the legal property of the latter. This grants labs absolute access to and control over the tissues, and an exclusive legal claim to patentable or commercial potential deriving from them (Waldby and Mitchell, 2006; Rajan 2007; Cooper 2008; Ikemoto 2009; Cooper and Waldby 2014; Parry and Greenhough 2018).

From fertility clinic to research lab

By identifying embryos seen to have no future life potential, preimplantation genetic tests operate to surplus embryos out of the reproductive stream and into the research lab. Once in the lab, these embryos become significant sources of biovalue: "a surplus value of vitality and instrumental knowledge which can be placed at the disposal of human subjects" (Waldby 2000, 19). Hierarchised as marginal tissue in the fertility clinic, in the lab these embryos can be transformed into technologies to aid in the intensification of vitality for other living beings. Preimplantation genetic tests thus precipitate a transmutation in the value of abnormal embryos: from future biographical life to raw biological resource.

This transmutation in value triggers a shift in understanding as they move from fertility clinic to research lab. In the clinic, the aim is to bring a baby to term. With the help of scientific and medical intervention, the embryo is set on a path of stable and progressive development in which human life promises to unfold steadily from the fertilised ovum. In the research lab, the embryo is set on a different path. Although similarly objective-driven, the goal is not the (re)production of a whole human body, of biographical life, but rather the cultivation of an active, manageable, flexible supply of unorganised tissue rich with potential for experimentation, invention and profit. The respatialisation introduced earlier is again a precondition here. Separated from their *milieu intérieur, ex vivo* embryos can circulate between scientists and researchers in labs around the world; their depths can be plumbed using a variety of technoscientific instruments.

Surplussed embryos thus bind the fertility industry to reproductive and regenerative medicine, biomedical research, tissue engineering and other emergent bioindustries (Franklin 2013). The transfer of these embryos from clinic to lab requires official relationships between those who have possession and control of the embryos in the fertility setting and those who would receive the embryos for research. Many labs have long-standing, institutional ethics review board-approved projects in association with one or more fertility clinics, which will send their way any embryos donated to research. "Usually yearly or so [clinics] will send out something to their patients saying 'do you want us to continue to preserve your embryos? Do you want us to discard them? Do you want us to donate them?' [These clinics] have that kind of process, and so they've been referring patients to us.... We get maybe 50 [embryos] per year. Generally patients will, when they're done, they'll just donate all of their remaining embryos," one researcher told me. Her lab had relationships with three major fertility clinics across southern California.

This kind of partnership is especially common between university research labs and their own, on-campus IVF centres: an easier and cheaper alternative to sourcing embryos from outside clinics. Some fertility clinics may have their own, in-house labs, in which they perform PGT and conduct biomedical research on surplussed embryos. At one clinic, I was treated to a tour of their state-of-the-art lab, where they were doing their own research on the accuracy of PGT maps. Sometimes, more than embryos travel. Lab directors may double as researchers, clinicians as stem cell scientists. In rare cases, an IVF specialist may run a fertility clinic, operate a human embryonic stem cell bank with lines derived from surplussed embryos and participate in stem cell research (Ikemoto 2009). These relationships evince the often-close connections between fertility clinics and research labs, connections that facilitate the use of IVF embryos in the laboratory setting.

A growing literature has examined the changing place of surplus embryos beyond the fertility clinic, with a particular focus on their use in regenerative medicine (Waldby 2002; Waldby and Squier 2003; Squier 2004; Franklin 2006; Waldby and Mitchell 2006; Cooper 2008; Svendsen and Koch 2008; Gottweis et al. 2009; Waldby and Cooper 2010; Cooper and Waldby 2014). This work has shown how embryonic vitality is repurposed in an effort to extend and renew existing human life, for example through stem cell thera-pies. While the immense medical possibilities touted by proponents of regen-erative medicine remain largely speculative, their effects are nonetheless material: from pressure on real-time life extension technologies like organ transplantation, to the investment of pension funds in the biotech sector, to high volumes of embryos flowing from fertility treatment to regenerative medicine (Waldby and Mitchell 2006).

If abnormal embryos are valuable in this context, it is largely in their potential capacity to produce normal cell lines (Lavon et al. 2008). Beyond the life sciences, however, scarce attention has been paid to the value of embryonic abnormality itself. Such an orientation reveals a somewhat differ-ent use for surplussed IVF embryos: as disease-in-a-dish models in the study of genetic and chromosomal mutation and pathological embryogenesis. Here, the cause of an embryo's devaluation in the fertility clinic becomes the source of its biovalue in the research lab. "Abnormal embryos are actually in some ways much more valuable for research than genetically normal ones," a human embryonic stem cell researcher told me.

Abnormal embryos' biovalue derives from their instrumentalisation as models of abnormal development. Diagnosed using preimplantation genetic tests, embryos with identified genetic and chromosomal abnormalities become indispensable models for research on early human development. They hold the promise of an unlimited supply of cells with which to study and map the mechanisms of disease more effectively than those provided by genetically altered or animal cells (Stephenson et al. 2013; see also Franklin 2013). Researchers have cultivated cell lines with genetic traits for Down syndrome, Huntington's disease, cystic fibrosis, BRCA, muscular dystrophy and a range of sex chromosome disorders, among other conditions.

Much as the anatomists of the eighteenth and nineteenth centuries had to "open up a few corpses" to better understand the organ-isation of disease (Foucault 1989, 152; Waldby 2000), researchers studying the genesis and development of abnormality have opened up a few embryos. Recall the quo-tation from earlier: "you can take an embryo, put it under a microscope, pluck out one cell, and just study it" (personal communication). In this way, the location and progression of abnormal development can be seen and examined, new biological maps created, old biological maps perfected. The knowledge generated by this research finds wide application, of course, in fertility practice, where it not only improves the precision and efficacy of preimplantation genetic tests, but also widens the array of characteristics subject to detection via PGT.

The derivation of biovalue from the embryos under consideration in this chapter is thus tied to their biological abnormality. But it is also tied to social understandings of the non-desirability of particular kinds of biological differ-ence, and to the construction of "abnormal" embryos as non-viable. As mentioned earlier, anecdotal evidence suggests that abnormal embryos are more likely to be donated to research, as opposed to being transferred or cryopreserved for future use. Stem cell researchers frequently describe these embryos as being free from the ethical constraints associated with using normal embryos in the lab (for example Stephenson et al. 2013; Huang et al. 2015). Having lost their ontological significance as the beginnings of human life, abnormal embryos become "matter out of place" (Douglas 1984) in the fertility clinic. As a result, they are more easily entered into the tissue economy. Classified as waste in one space, they can become the starting point for the generation of new kinds of value in another. These tissues become bioavailable, then, in part through the setting up of a bio-social hierarchy that deems some embryos more valuable than others in the clinical context.

Conclusion

While a primary focus in the fertility clinic, in reproductive medicine and in biomedical research, abnormal embryos have garnered scant attention beyond the life sciences. And yet they are deeply social and political entities, deserving of careful and critical consideration by social scientists. An overarching aim of this chapter has been to contest the idea that embryonic abnormality is a pre-given, biological category. Produced at the intersection of reproductive desire, mapping techniques, laboratory technology, bio-social norms and human biology, the abnormal IVF embryo is less detected than made. Decisions about which embryos to transfer and which to discard are over-determined by socially, temporally and geographically contextual concepts of idealised form and function. As such, I argue, they reflect a hierarchy of human life that values particular kinds of biological difference over others.

I am treading uncomfortable territory here. Inferring the differential valuation of human life from embryo selection decisions risks reifying embryos as future persons. I find it absolutely essential to resist and dismantle any such reification. I am not opposed to the use of ARTs to prevent severe illness, impairment and suffering on behalf of a future child or their parents, although it is imperative that we unpack what we mean by "severe", "illness", "impairment" and "suffering" and that we question the disciplinary effects of biomedical authority. I unequivocally support a woman's right to terminate a pregnancy as a crucial dimension of a reproductive politics that also supports a woman's right to carry a pregnancy to term and to be cared for while doing so. And I am not against the use of embryos in biomedical research, although I join those concerned with how these embryos are procured and from whom.

Like cancer, skin or blood cells, embryos are a source of regenerative genetic material with significant biovalue. But unlike these other cell types, embryos have the potential to produce a human being. Indeed, as Sarah Franklin has argued (2006), embryos are the objects of scientific investment for the same reason that they are the objects of significant moral and social investment: because they have the potential to be biologically reproductive. This potential matters, discursively and materially. It complicates how we think about, imagine and deal with both *in* and *ex vivo* embryos as political, social, legal, economic and biomedical entities. And it challenges us to think about how we understand, order and reproduce life in a biotech age.

Notes

1 Lawrence Cohen (2005, 83) writes, "to be bioavailable in my terms is to be available for the selective disaggregation of one's cells or tissues and their reincorporation into another body (or machine)".
2 Practitioners call this practice intentional diminishment. Although cases of intentional diminishment are exceedingly rare, they are a frequent topic of discussion in the field of assisted reproduction. Many see it as a misuse of technology intended to de-select, not reproduce, illness and disability.
3 BRCA stands for BReast CAncer susceptibility gene. The existence of this gene gained popular awareness when, in 2013, Angelina Jolie published an opinion piece in the *New York Times* entitled "My Medical Choice". In it, she chronicled her preventative double mastectomy, intended to reduce her likelihood of developing breast cancer as a carrier of the BRCA gene.
4 Carrier genetic testing is a form of personalised genetic screening in which carriers— intended parents or gamete donors—are tested for heritable genetic conditions.
5 Because cryopreservation and donation can be costly, these embryos are often simply discarded. Clinicians spoke with thinly veiled frustration about a patient population who "wasted" (i.e., declined to transfer) normal embryos of the undesired sex.
6 In reality the process is anything but ineluctable, "not a simple process of steps leading to potential success", but a "confusing and stressful world of disjointed temporalities, jangled emotions, difficult decisions, unfamiliar procedures, medical jargon, and metabolic chaos" (Franklin 2013, 7; see also Franklin 1997).
7 Down syndrome and sex chromosome abnormalities (e.g., Kleinfelter's syndrome and Turner's syndrome) are commonly cited as examples.
8 Some clinics will not transfer abnormal embryos under any circumstances. If intended parents want to proceed with the transfer, they have to change clinics.
9 Edwards made this statement during his comments to the 1999 meetings of the European Society of Reproduction and Embryology. Lois Rogers, "Having Disabled Babies Will Be 'Sin', Says Scientist," *Sunday Times* (London), 4 July 1999. Quoted in Kafer 2013, 164–165.

References

Canguilhem, Georges. 1991. *The Normal and the Pathological*. New York: Zone Books.
Cohen, Lawrence. 2005. "Operability, Bioavailability, and Exception". In *Global Assemblages: Technology, Politics, and Ethics as Anthropological Problems*, edited by Aihwa Ong and Stephen J. Collier, 79–90. London: Blackwell Publishing.
Cooper, Melinda. 2008. *Life as Surplus: Biotechnology and Capitalism in the Neoliberal Era*. Seattle: University of Washington Press.
Cooper, Melinda, and Catherine Waldby. 2014. *Clinical Labor: Tissue Donors and Research Subjects in the Global Bioeconomy*. Durham: Duke University Press.
Cromer, Risa. 2017. "Waiting: The Redemption of Frozen Embryos through Embryo Adoption and Stem Cell Research in the United States." In *The Anthropology of the Fetus*, edited by Sally Han, Tracy K. Betsinger and Amy B. Scott, 171–199. New York and Oxford: Berghahn Books.

Douglas, M. 1984. *Purity and Danger: An Analysis of the Concepts of Pollution and Taboo.* London: Routledge.

Duden, Barbara. 1993. *Disembodying Women: Perspectives on Pregnancy and the Unborn.* Cambridge: Harvard University Press.

Foucault, Michel. 1989. *The Birth of the Clinic.* Oxon: Routledge.

Foucault, Michel. 2003. *Abnormal: Lectures at the Collège de France 1974–1975.* New York: Picador.

Franklin, Sarah. 1997. *Embodied Progress: A Cultural Account of Reproduction.* London: Routledge.

Franklin, Sarah. 2006. "Embryonic Economies: The Double Reproductive Value of Stem Cells." *Biosocieties 1(1):* 167–188.

Franklin, Sarah. 2013. *Biological Relatives: IVF, Stem Cells, and the Future of Kinship.* Durham: Duke University Press.

Franklin, Sarah, Celia Roberts, Karen Throsby, Peter Braude, J. Shaw and Alison Lashwood. 2005. "Factors Affecting PGD Patients' Consent to Donate Embryos to Stem Cell Research." Paper presented at Sixth International Symposium on Pre-implantation Genetics, London, 19–21 May.

Gordon, Ian R. 2003. *Laboratory Production of Cattle Embryos,* 2nd ed. Wallingford, U.K.: CABI.

Gottweis, H., B. Salter and C. Waldby. 2009. *The Global Politics of Human Embryonic Stem Cell Science: Regenerative Medicine in Transition.* Basingstoke: Palgrave.

Greenhough, Beth, Bronwyn Parry, Isabel Dyck, and Tim Brown. 2015. "Introduction: The Gendered Geographies of 'Bodies Across Borders'." *Gender, Place & Culture: A Journal of Feminist Geography* 22(1): 83–89.

Hall, Edward. 2003. "Reading Maps of Genes: Interpreting the Spatiality of Genetic Knowledge." *Health & Place 9:* 151–161.

Haraway, Donna. 1997. *Modest_witness@second_millenium.femaleman_meets_Onco-Mouse: Feminism and Technoscience.* New York and London: Routledge.

Huang, B. et al. 2015. "Establishment of Human-Embryonic-Stem-Cell Line from Mosaic Trisomy 9 Embryo." *Taiwanese Journal of Obstetrics & Gynecology54(5):* 505–511.

Ikemoto, Lisa. 2009. "Eggs as Capital: Human Egg Procurement in the Fertility Industry and the Stem Cell Research Enterprise." *Signs 34(4):* 763–781.

Jaroff, Leon. 1989. "The Gene Hunt". *Time,* 10 March, 1989. http://content.time.com/time/magazine/article/0,9171,957263,00.html

Jolie, Angelina. 2013. "My Medical Choice." *New York Times,* 14 May 2013. http://www.nytimes.com/2013/05/14/opinion/my-medical-choice.html

Kafer, Alison. 2013. *Feminist, Queer, Crip.* Bloomington: Indiana University Press.

Kayayerli, Damla. 2017. "Genome Project Promises Super Olympic Athletes." *Daily Sabah Feature,* June 13, 2017. https://www.dailysabah.com/feature/2017/06/13/genome-project- promises-super-olympic-athletes

Kukla, Rebecca. 2005. *Mass Hysteria: Medicine, Culture, and Mothers' Bodies.* Lanham: Rowman & Littlefield.

Landecker, Hannah. 2005. "Living Differently in Time: Plasticity, Temporality and Cellular Biotechnologies." *Culture and Machine, 7,* an e-journal.

Landecker, Hannah. 2006. *Culturing Life: How Cells Became Technologies.* Harvard University Press: Cambridge, MA.

Lavon, Neta, Kavita Narwani, Tamar Golan-Lev, Nicole Buehler, David Hill and Nissim Benvenistry. 2008. "Derivation of Euploid Human Embryonic Stem Cells from Aneuploidy Embryos." *Stem Cells 26:* 1872–1882.

Lewis, Sophie. 2016. "Gestational Labors: Care Politics and Surrogates' Struggle." In *Intimate Economies: Bodies, Emotions and Sexualities on the Global Market*, edited by Susanne Hofmann and Adi Moreno. New York: Palgrave.

Lewis, Sophie. 2017. "Defending Intimacy Against What? Limits of Antisurrogacy Feminisms." *Signs: A Journal of Women in Culture and Society 43(1)*: 97–125.

Lippman, Abby. 1991. "Prenatal Testing and Screening: Constructing Needs and Reinforcing Inequities." *American Journal of Law and Medicine 17(1–2)*: 15–50.

Mansfield, Becky. 2012. "Gendered Biopolitics of Public Health: Regulation and Discipline in Seafood Consumption Advisories." *Environment and Planning D: Society and Space 30(4)*: 588–602.

McWhorter, Ladelle. 2009. *Racism and Sexual Oppression in Anglo-America: A Genealogy*. Bloomington: Indiana University Press.

McRuer, Robert. 2006. *Crip Theory: Cultural Signs of Queerness and Disability*. New York: NYU Press.

Morgan, Lynn. 2009. *Icons of Life: A Cultural History of Human Embryos*. Oakland: University of California Press.

Mukherjee, Siddhartha. 2016. *The Gene: An Intimate History*. New York: Scribner.

Oakley, Ann. 1984. *Captured Womb: A History of the Medican Care of Pregnant Women*. New York: Basil Blackwell Publishing Ltd.

Ordover, Nancy. 2003. *American Eugenics: Race, Queer Anatomy, and the Science of Nationalism*. Minneapolis and London: University of Minnesota Press.

Parry, Bronwyn. 2012. "Economies of Bodily Commodification." In *The Wiley-Blackwell Companion to Economic Geography*, edited by Trevor Barnes, Jamie Peck, and Eric Sheppard, 213–225. Malden, MA: Blackwell.

Parry, Bronwyn and Beth Greenhough. 2018. *Bioinformation*. Cambridge: Polity Press.

Parry, Sarah. 2006. "(Re)constructing Embryos in Stem Cell Res: Exploring the Meaning of Embryos for People Involved in Fertility Treatments." *Social Sciences & Medicine 62*: 2349–2359.

Payne, Jenny Gunnarson. 2015. "Reproduction in Transition: Cross-Border Egg Donation, Biodesirability and New Reproductive Subjectivities on the European Fertility Market." *Gender, Place & Culture: A Journal of Feminist Geography 22(1)*: 107–122.

Philo, Chris. 2000. "The Birth of The Clinic: An Unknown Work of Medical Geography." *Area 32(1)*: 11–19.

Rajan, Kaushik Sunder. 2007. *Biocapital: The Constitution of Post-Genomic Life*. Durham: Duke University Press.

Rapp, Rayna. 1999. *Testing Women, Testing the Fetus: The Social Impact of Amniocentesis in America*. New York: Routledge.

Regalado, Antonio. 2017. "Eugenics 2.0: We're at the Dawn of Choosing Embryos by Health, Height, and More." *MIT Technology Review*, 1 November 2017. https://www.technologyreview.com/s/609204/eugenics-20-were-at-the-dawn-of- choosing-embryos-by-health-height-and-more/

Schurr, Carolin. 2016. "From Biopolitics to Bioeconomies: The ART of (Re-)Producing white Futures in Mexico's Surrogacy Market." *Environment and Planning D: Society and Space 35(2)*: 241–262.

Schurr, Carolin and Elisabeth Militz. 2018. "The Affective Economy of Transnational Surrogacy." *Environment and Planning A: Economy and Space* https://doi.org/10.1177/0308518X18769652.

Squier, Susan. 2004. *Liminal Lives: Imagining the Human at the Frontiers of Biomedicine*. Durham: Duke University Press.

Stephenson, Emma, Dusko Ilic, Laureen Jacquet, Olga Genbacev, and Susan J. Fisher. 2013. "PGD and Human Embryonic Stem Cell Technology." In *Preimplantation Genetic Diagnosis in Clinical Practice*, edited by T. El-Toukhy and Peter Braude, 153–164. London: Springer.

Svendsen, Mette N. and Lene Koch. 2008. "Unpacking the 'Spare' Embryo: Facilitating Stem Cell Research in a Moral Landscape." *Social Studies of Science 38(1)*: 93–110.

Thompson, Charis. 2013. "Governance, Regulation, and Control: Public Participation." In International Summit on Human Gene Editing: A Global Discussion, 1–3 December 2015, Washington, DC. http://eprints.lse.ac.uk/66585/

Tremain, Shelley. 2006. "Reproductive Freedom, Self-Regulation, and the Government of Impairment in Utero." *Hypatia 21*: 35–53.

Wald, Priscilla. 2000. "Future Perfect: Grammar, Genes, and Geography." *New Literary History 31(4)*: 681–708.

Waldby, Catherine. 2000. *The Visible Human Project: Informatic Bodies and Post-Human Medicine*. London: Routledge.

Waldby, Catherine. 2002. "Stem Cells, Tissue Culture, and the Production of Biovalue." *Health: An Interdisciplinary Journal for the Social Study of Health, Illness, and Medicine 6(3)*: 305–323.

Waldby, Catherine and Melinda Cooper. 2010. "From Reproductive Work to Regenerative Labour: The Female Body and The Stem Cell Industries." *Feminist theory 11(1)*: 3–22.

Waldby, Catherine and Robert Mitchell. 2006. *Tissue Economies: Blood, Organs, and Cell Lines in Late Capitalism*. Durham: Duke University Press.

Waldby, Catherine and Susan Squier. 2003. "Ontogeny, Ontology, and Phylogeny: Embryonic Life and Stem Cell Technologies." *Configurations 11*: 27–46.

Zwart, Hub. 2009. "Genomics Metaphors and Genetic Determinism." In *New Visions of Nature: Complexity and Authenticity*, edited by Martin Drenthen, Jozef Keulartz, James Proctor, 155–172. London: Springer.

3   Right donor, right place
Spatialities of artificial insemination

Marcia R. England

Introduction

Some people think conception is as easy as sperm + egg = pregnancy. But for many, it's not quite that simple. Fertility issues and/or lack of access to fresh sperm complicate that equation. In "traditional" insemination, the sperm can live for up to five days in a woman's body in order to meet—and possibly fertilise—the egg. Those who use frozen sperm have a window of 12–24 hours (the lifespan of frozen sperm) to get it right. Timing that correctly with ovulation (approximately a 24-hour window) involves precise calculation. Women who are actively trying to get pregnant become familiar with almost all aspects of their bodies from cervical mucus to ovulation pains to anticipated implantation to either menses or a BFP—big fat positive pregnancy test. They urinate on test sticks to track ovulation and possible pregnancy. Some refer to themselves as POAS (pee on a stick) addicts, checking urine as many as six times a day. They measure basal body temperature (BBT) and meticulously pore over their BBT charts to pinpoint ovulation and possible implantation. These aspects of observing and tracking one's menstrual and ovulatory cycles complicate the idea that pregnancy just "happens". For many, pregnancy eludes them due to a number of circumstances (e.g., infertility for one or both partners or lack of access to sperm) and they seek alternatives to "traditional" insemination in order to conceive.

This project explores reproductive geographies by interviewing clients of a cryobank in the Pacific Northwest of the United States. Women and couples (both same sex and heterosexual) purchase sperm from the cryobank for artificial insemination (AI) either at home or at a clinic. In this chapter, I examine 1) how clients make decisions regarding sperm donor and location of insemination and 2) the mobilities of cryobanks. The multiple modes of sperm "delivery" show just one example of the geographies of artificial insemination. Clients are given a wide array of information regarding the physical and social characteristics of donors in order to aid in their donor decision. These decisions demonstrate how sperm can be imagined to deliver desired traits of donors. When the choice of donor is made, there is an additional geography to be explored—that of the site of insemination. Some find the process to be intimate and prefer the private space of the home, while others go the clinic route.

Reproduction has a number of spatialities. Often intimate and sometimes hidden, conception is geographical in many ways. Fertility is emerging as a critical question in modern society as infertility is on the rise and new technologies have been utilised in order to address this issue. Geographers who research "artifical" means of conception often engage with social and medical geographies (e.g., reproductive health and human rights) as socio-spatial concerns are intimately connected to bodies. Below, I explore the associated spatialities of artificial insemination including geographies of the body and mobilities. In addition, I discuss how understandings of intimacy and place relate to sites of insemination. This work seeks to understand how geographies of the body and (potentially) intimate medical geographies play out in reproductive decision-making. This research adds to geographical literature in a novel way. It discusses the role of fantasy, privacy and intimacy in reproductive decisions and addresses how understandings of place inform those choices.

Geographies of the body

Geographical research on the body explores many topics, but reproductive geography is a newly emerging subfield. Reproductive geography connects socio-cultural and medical geography (Boyer 2010; Fannin 2013; Colls and Fannin 2013). This research seeks to expand the literature in these areas by addressing how decisions made in the growing phenomenon of artificial insemination create complex geographies of the body and influence understandings of what can be called "geographies of intimacy" (see Valentine 2008) and private space.

Foucault, in *The Birth of the Clinic* (1973), describes the body as a geographical system. Philo's (2000, 13) exploration of Foucault's book discusses "a thoroughly spatial system of bones, organs, blood vessels, fatty deposits and tissues. Attention duly shifted to this 'geographical system' that was the body itself". Within this chapter, I hope to demonstrate that the body is an important site in geography and essential when discussing reproduction and reproductive decisions.

Geographers have explored the subject of the body in many ways, but much geographical research on the body deals with the superficialities of the body: body size, ability, attractiveness, adornment (e.g., Longhurst 1997, 2004, 2012; Colls 2006; Longhurst and Johnston 2014). While the exterior of the body plays an important role in donor decisions, this research contributes to a new direction in geography by focusing on bodily interiors and conception through examination of artificial insemination. Emily Martin (2001, 20) discusses the fragmentation of the body that can occur in reproductive technologies:

> Human eggs, sperm, and embryos can now be moved from body to body or out of and back into the same female body. The organic unity of fetus and mother can no longer be assumed, and all these newly fragmented

parts can now be subjected to market forces, ordered, produced, bought, and sold.

This fragmentation, and subsequent mobility, of body parts is key to understanding the changing landscape of reproduction. New technologies allow people to conceive who historically may not been able to get pregnant and/or bring that pregnancy to term.

Eugenics, elitism, and exclusion

Using donated gametes to conceive creates new geographies of bodies due to the choices involved. These decisions can be based on a number of factors, but they are rooted in the desire for certain traits in one's future child. The issue of eugenics is often raised when discussing donor selection. Eugenics is the method of "improving" humans through controlled and selective mating. Cynthia Daniels and Janet Golden (2004, 6) argue that the purchase of sperm is a form of eugenics:

> Purchasers of "procreative compounds" can select sperm that comes from donors who match partners or from those who embody idealized versions of men—taller, handsomer, smarter than the average. Vials of sperm selected by purchasers, like sperm donors once selected by doctors contain, figuratively, the cultural ranking of particular traits. Selections are made, in part, on the basis of nonheritable characteristics and so reveal prevailing social hierarchies and the operation of what we call populist market eugenics. We define this as the belief that certain human traits can be "purchased" through the careful selection of sperm. We note that the traits which appear to be most in demand in the contemporary sperm banking industry reflect not just the desires of consumers to have offspring who physically resemble them, but to have offspring who will be at the top of prevailing social hierarchies.

The purchase of donor sperm is, for some, essentially the purchase of desired traits. There is an element of fantasy to explore here as well. The donor is imagined/fantasised to look a certain way, carry specific traits, be what the client wishes him to be.

Tied to this are marketing materials for cryobanks. They advertise an idealised donor to clients. For example, the Fairfax Cryobank (not the cryobank researched in this chapter) offers new ways to search for the "perfect sperm donor":

> Finding the perfect sperm donor just got easier, thanks to Fairfax Cryobank's new donor search tool, which allows potential clients to search by lifestyle criteria (including astrological signs, favorite songs, life goals, favorite pets and favorite subjects in school), as well as by traditional

physical and educational standards. The search tool's "Basic" option allows potential users to search by donor type, ancestry, blood type and phenotype. The "Advanced" search allows clients to search by ethnicity, skin tone, CMV status,[1] specimen type or photos. The new "Lifestyle" option allows potential users to search based on the sperm donor's personal preferences and other attributes, including astrological sign, religion, favorite pet (dogs, cats, fish, bird, etc.), favorite subject, personal goals (fame, community service, travel and others) and hobbies.

Here we see Daniels and Golden (2004) exemplified. The above advertisement touts the idea that desired human characteristics can be "purchased" with the help of a search function. Clients select donors with desired traits as those characteristics are often imagined as embedded within sperm. Following this, Diane Tober (2001, 138) discusses how desired traits affect choice:

> The sperm-banking industry and the market for sperm are both heavily influenced by the notion that some traits—social or physical—are more desirable than others, and that these traits reside in the sperm. Semen is a vehicle for the transmission of genetic material; as such, various complex meanings—biological, evolutionary, historical, cultural, political, technological, sexual—intersect at this particular site.

The selection of a donor is a complicated process and it is difficult to assess whether it is eugenics or just consumer preference. Bronwyn Parry (2015) argues that donor profiling by cryobanks of whom they receive specimens from may be a sorting of "good" versus "bad" stock with implications for improving "population quality". An example of tension in the preference-eugenics debate may be red-haired donors. Cryobanks used to reject red-haired men from donating because they were not profitable (D'Arcy 2011). Now there is a demand (Tourielle 2017).

Mobility

Mobility plays a key role in artificial insemination. The study of mobility and mobilities in geography is a rapidly growing field but has—as Tim Cresswell and Peter Merriman (2011) argue—always been a part of geography. Mobilities happen at many levels—body movements, circulation, transportation, social interactions and so on (see Cresswell 2006; Urry 2007; Cresswell and Merriman 2011). There are at least three scales of mobility in artificial insemination: the sperm within the semen, the shipping of the vials of sperm sold and the potential travel of clients to clinics or other insemination destinations. These three separate, but related, factors of artificial insemination show geographical dimensions of reproduction.

The mobility of sperm is often discussed in detail in reproduction. Often referred to as "swimmers", the calculation of how long it will take for the sperm to meet the egg often fascinates those trying to conceive at home—does it take six or 12 hours for the sperm to bypass the cervix? How long will the sperm need to make it up the fallopian tube? How many hours after the LH surge[2] will ovulation occur? Twelve, 24 and/or 36 hours after the LH surge are popular times for those who do intracervical insemination (ICI) at home. When performing an intrauterine insemination (IUI) at a clinic, most doctors will wait 24–36 hours after a trigger shot.[3] The trigger shot is an hCG (human chorionic gonadotropin, which is often referred to as the pregnancy hormone as pregnancy tests measure hCG levels) injection administered to the stomach that controls timing of ovulation better.

Higher sperm counts with high motility have higher chances of fertilising an egg. Motility is the forward, swimming motion of sperm. For a sperm to fertilise an egg, it must be able to move rapidly through the female reproductive system, which requires strong swimming action. The World Health Organisation (WHO) defines normal sperm motility as 32 per cent of the total sperm having progressive motility (WHO 2010).[4] In order to be "normal", more than 40 million motile sperm should be present in the ejaculate.

The movement of sperm (in dewars) from cryobank to client is another spatial aspect of artificial insemination. While Ruth Chadwick and Alan O'Connor (2015) discuss movement in an international context, the mobility of the dewar as a container of gametes is also important to consider. Because of technology, gamete mobility is possible. The commodified sperm specimen is now mobile. It is, as per Martin's (2001) argument mentioned earlier in this chapter, fragmented and divorced from the body. As such, its mobility is increased.

The third form of mobility is client travel in order to inseminate. It can be a form of fertility/reproductive tourism, which is travel to another state or country in order to save money on procedures/drugs or to procure reproductive materials (Bergmann 2011). Marcia Inhorn and Pasquale Patrizio (2009) describe it as "reproductive exile" in that the travel is costly, stressful and usually done only out of necessity. While client travel may be more localised when visiting a reproductive endocrinology clinic, mobility is still necessary. Without mobility, the client may not have the means to attempt conception.

Sperm donation

Below, I discuss the use of frozen sperm for artificial insemination and the processes that surround the insemination of said sperm. Rene Almeling (2011) observes that fresh sperm went out of vogue during the 1980s AIDS epidemic. Following the scare, standard procedure for cryobanks was to test, freeze and quarantine (for six months) donated sperm. Still standard today, quarantining of sperm helps to reduce transmission risks of sexually transmitted infections (STIs).

The United States dominates the emerging international market in human sperm (Parry 2015). Embryo, oocyte (egg) and sperm donations in the U.S. are often anonymous, although they can also be open or known.[5] While in other countries (such as Australia, Germany, the Netherlands, the United

Kingdom and Norway) anonymous donation is illegal, there is no such law in the United States. Anonymity can be assured in the U.S.[6] Those countries that prohibit anonymous donation do so in order to secure the child's right to biological knowledge about their parentage.

Process

Artificial insemination is a process by which sperm is deposited into the vagina next to the cervix (intracervical) or by crossing the cervix into the uterus (intrauterine) shortly before ovulation. Timing depends on method of insemination and often age. For intracervical insemination (ICI), the timing typically ranges from 12–36 hours after the LH surge, while intrauterine insemination can be performed 24–36 hours after the hormone surge is either detected or induced with hormonal injections. In the case of IUI, sperm is "washed" in a laboratory in order to eliminate irritating chemicals in semen. A chemical is added to separate the most motile sperm. The specimen is then spun in a centrifuge to get the most active sperm. Sperm, when washed, have a greater likelihood of locating an egg. Sperm are then placed via catheter into the uterus either by a doctor in a clinic or in a home setting. ICI does not require the washing process because the semen does not bypass the cervix.

In a doctor-assisted IUI, clomiphene or letrozole is often prescribed (to induce ovulation and stimulate follicles and eggs) alongside a trigger shot, which takes the place of a natural LH surge. Doctors use hormonal injections to trigger ovulation and release of egg(s). In addition to the drugs prescribed, ultrasounds are often used to track ovulation and count the number of follicles/eggs. Some doctors will let the client track her surge naturally and put the onus of scheduling insemination on her.

The calculations and precise timing needed in artificial insemination demonstrate that the concept of "natural" may very well be meaningless in this situation. Bodies are managed and manipulated in order to achieve pregnancy and to carry that pregnancy to term. Fertility and reproduction as a simple process is disrupted in the world of artificial insemination.

The cryobank

The cryobank examined for this case study is accessible to clients who wish to purchase sperm and inseminate without medical intervention. It was one of the only cryobanks in the U.S. that delivered to a home address and did not require doctor supervision. Other cryobanks require that the dewar containing the specimen(s) be delivered directly to a physician's office or clinic. The vast majority of cryobanks in the United States require medical supervision, meaning that a doctor is aware of the patient's choice to use frozen sperm. Cryobanks often require no more than a signature from the doctor. A doctor, however, may choose to be more involved. Some require extensive psychological and physiological testing of the patient to be inseminated.

This cryobank was relatively inexpensive compared to other cryobanks in the U.S. ($275–$375 per cc).[7] Other cryobanks charge $500–$900. However, the associated costs add up quickly (e.g., insemination kits are $10–25, shipping is an additional $150–$300 depending on speed of delivery). This is just for at home insemination. Using a reproductive endocrinologist (RE) can add hundreds or even thousands more between prescriptions, ultrasounds and the actual insemination.

Methods

Participants in this research were recruited via a public forum on conception, pregnancy and parenting associated with the cryobank, as well as through the snowball method. The female-only respondents were a mix of women involved in heterosexual (HC) and lesbian (LC) couples and women seeking to be a Single Mother by Choice (SMBC). Women were interviewed via the forum, email and phone calls. Each participant was asked to describe her decision-making processes in the choice of sperm donor (those who used embryo adoption or egg donors were not included) and insemination site. Only those who performed intracervical and intrauterine inseminations were considered. Additional data such as age and relationship status were also collected. The women ranged in age from mid-twenties to late forties. No economic or ethnic/racial information was collected.

To demonstrate the spatialities of artificial insemination, I use materials from the cryobank, the cryobank's online forum and participant responses to frame my analysis. First, I discuss the relationship between bodies and artificial insemination. Second, I discuss the role of intimacy in choosing sites of insemination and third, I focus on the mobile aspects of this form of reproduction.

Bodies and artificial insemination

Donating bodies

The American Society for Reproductive Medicine, which makes policy recommendations, advocates that payment for sperm donation should not be so lucrative that it may constitute the primary motivation for donation. Prices are set by each bank and range from $50 to $200. For the cryobank examined in this chapter, donors may earn as much as $1000 monthly with a six to 12-month commitment required. Donations are made two to three times a week.

Mette Svendsen (2007) explores why donors choose to donate in a discussion of "intimate citizenship" and whether the decision is made based on a contribution to society. Recruitment for donation often centers around altruism and making a difference in infertile women's/couple's lives. Materials for the cryobank researched here state: "Donating your sperm is a caring and generous act. Those couples that use donor sperm feel deep gratitude and respect for the gift you give so willingly".

There is also an elitism to the donation process, which leads us back to the issue of eugenics. From the cryobank's website, the minimum qualifications are:

> If you are at least 5'10", between the ages of 18 and 35 years, have post-secondary education (college, vocational or technical) and are within normal limits of weight for your muscular build and height, you may be a potential candidate to become a donor. We need donors with all types of racial and ethnic backgrounds. We are not able to accept applicants who use tobacco products in any form, including smoking or ingesting marijuana. Use of these substances will be tested for throughout the donation process.

The qualifying traits of the donors indicate idealised characteristics. They demonstrate that only certain bodies are valued by science, by society, by clients.

Receiving bodies

In anticipation of insemination, participants watched and prepared their bodies—sometimes for months or years. Most tracked their basal body temperature or monitored their cycle with ovulation kits. Those that did this nearly exclusively used an online site or an application on their phones to track the data they collected. This data included tracking their menstrual cycle, cervical mucus and temperature. All took prenatal vitamins, which specifically included folic acid and DHA.

Participants' at home pre-insemination preparations included drinking pomegranate or grapefruit juice (to supposedly thicken the uterine lining or thin cervical mucus respectively) and eating the core of a pineapple. Fresh pineapple contains bromelain which can act as a blood thinner and can increase blood flow to the uterus. Pineapple also contains selenium, which helps to thicken the uterine lining and so can aid implantation. Other foods touted as fertility enhancers include: Brazil nuts (thickens uterine lining), maca (for egg health and increased progesterone), royal jelly (for egg health and quality), full fat dairy (decreases fertility issues). Foods often avoided are: non-organic bananas, caffeine, alcohol and non-pasteurised cheeses. Another commonly used product to thin cervical mucus is cough syrup. Participants enthusiastically discussed their body preparations:

> I started taking Royal Jelly…and this cycle I ovulated a couple of days earlier. Last month was a bad long month for me so it is great that this month is back on track.
>
> (Bea, 32, HC)[8]

> I took Mucinex starting two to three days before ovulation. You don't need to continue after [insemination] because the idea is to facilitate passage for the swimmers. Once the race is over, Mucinex's work is done!
>
> (Wren, 35, SMBC)

Interesting to note here is the desire to control "natural" processes. Women in this research manipulate their bodies both through foods and drugs in order to give themselves the best chance. While this is understandable considering the expense involved and the high desire for children, it also can be a moment in which the dualism of women and nature is considered for these practices both destabilise and reinforce this dualism. The nature–science binary has long associated women with the natural and tied women to their bodies due to processes such as menstruation, gestation, childbirth and lactation (Rose 1993; McDowell 1999). Think, for example, of the term "Mother Nature", which links women to nature and their childbearing capabilities.

For some, the process of trying to conceive is all-encompassing. It can dictate every aspect of one's life and, frankly, be consuming. Monitoring of bodily functions and diet can lead to a constant awareness of the processes involved. The body can become a site of obsession, in which the body's only purpose can be framed as reproductive in extreme cases.

> My obsession with implantation has led me back to researching the time frame for it. Implantation usually happens on average nine days past ovulation but can be six to 12 days later A positive urine test happens four to five days after implantation. So, if average is 9 + 5 more days = 14 days for positive. Of course, that depends on the woman. Even armed with this information I still POAS at eight days past ovulation. What is wrong with me? Of course it was negative. Last month I was away at a conference during the two week wait and I still obsessed about testing.
>
> (Tyra, 32, SMBC)

During AI and other assisted reproductive technologies, the hormones and procedures involved are taxing on the body and mental state (Woliver 2002; Blake et al. 2014). To complicate matters, women suffering from infertility face increased rates of depression and anxiety (American Pregnancy Association 2017). While critiqued for small sample sizes and over-reliance on treatment seekers, studies show that guilt, helplessness, marital stress and depression are common for those experiencing infertility (Culley et al. 2009). Some women note that being around pregnant women fills them with jealousy and/or depression. Many women in this research suffered damaging psychological effects during the endeavour of trying to conceive.

> I've been in a severe depression for days. I mean…serious depression… over the thought that I will never have a baby. I don't know how to face that or what to do with my life otherwise. It's immense grief over the fact that every time I start my period I feel like I've lost the child I've been believing was coming for the *two* weeks prior and I don't know how to make myself not think that way.
>
> (Mona, 41, SMBC)

Choosing a donor

When choosing a donor, there is excitement mixed with nerves. He is part imagination and fantasy, part pedigree. Clients often wish to know as much as they can about their donor in order to find the "right" match. In the donor application, the cryobank states:

> Clients normally desire to "know" the donor through the information you will provide in this history. If children are told that they were conceived through the use of a sperm donor, the information contained within this document may be very important to them for medical and psychological reasons.

The profiles the donors fill out are very important in the decision-making process. Basic donor profiles, which are free, contain information about certain traits such as interests, communication status, staff impressions about personality and physical features (eye colour, eyebrow shape and thickness, hair colour and type and eye and ear shape), ethnicity, blood type, height, build, vision, dentition, allergies, alcohol and tobacco usage, basic family health, education and CMV status. A short, handwritten essay is often included.

The extended profiles (usually over 50 pages of information) provided by donors are sometimes purchased[9] by cryobank clients in order to have a more in-depth medical history and know a bit more about the interests and hobbies of the donor. The extended version has details such as education status and career objectives, family education and accomplishments, ethnic origins, family physical characteristics, extensive family health histories of extended family (including mental health, visual and olfactory issues, cancer, blood issues, gastrointestinal health, respiratory problems, et cetera), a description of childhood years and memories, religion, activities and self-described personality characteristics. Dietary supplements and caffeine usage are even documented. These very detailed pages help to connect the client to the donor. They help to supplement the picture of the imagined donor through insight into health, hobbies and personality, as explained by one of my participants:

> As far as the extended profile, the part I found most useful was the extended family history. There are also long-form questions that the donor answers, some of which seemed kinda silly to me (i.e., favourite color and favourite song), but it is good for getting a better idea of the donor's writing style and by extension, I think, personality. I chose the donor first, then bought the extended profile and photos [also available for purchase] and felt good about my choice.
>
> (Jenn, 37, SMBC)

The decisions involved in choosing donors are based on a number of factors, including physical characteristics, blood type and education. The following quotations show how individual participants made decisions regarding sperm donors.

> Since I found no donor on the website that really perked my interest, I used a point system. A little OCD [obsessive compulsive disorder] I guess, but I thought it gave me the best way to objectively pick a donor. I start every donor off at a 3.0. I would add or deduct .25 points depending on major stuff like medical history, personality, et cetera. Some stuff like handiness or music talent I considered a neutral issue and it received an addition of .25 if it was a trait I wanted, but they didn't lose any points by not having the trait. I went through everything on the profile and even put a score to the writing portion. I ended up picking the donor that had the highest point total. When I actually describe the process, it sounds totally crazy although I feel I was able to compare the donors fairly and holistically using this method.
>
> (Rosanna, 38, SMBC)

> One of my biggest things was healthy personal/family history. So, I chose someone who had a similar family [history] as myself. I wanted someone who either had the same blood type as myself or who was type O like other members of my family. I wanted to have higher chance of being a donor if my child had health issues. My original choice sold out, but I was able to purchase another member's vials and he matched my requirements.
>
> (Becky, 27, SMBC)

Clients have a wealth of information available to them while making their choices. For some, likeness to a particular individual can be important when choosing a donor. Some clients want a donor who resembles them, their partner, another donor (for sibling purposes) or even someone famous.

While childhood photos may be available for purchase, clients have no idea what the adult donor looks like. While facial/photo matching services exist at the cryobank studied here, it only matches features such as eye shape (not colour), face shape and complexion. Hair colour or type, ethnicity and blood type are those characteristics not considered, but those notes are available when browsing donor profiles.

> My wife and I are brand new to this and finding it very difficult to choose a donor. We feel very uncomfortable having no idea what these men look like. I am not a shallow person, but I think it's fair to say that most people, if given the choice, would like to have a child with someone that they would deem physically attractive. Five feet eleven [inches], blonde hair, and blue eyes just doesn't tell us much. Some banks offer baby photos, but I just think in many cases grownups look nothing like they did when they were a baby. Most babies are cute by nature and some of

them grow up to be not-so-cute adults. Only one bank that I know of offers adult photos and they are very expensive. We would just like to have a better visual of this guy's actual features.

(Cody, 33, LC)

Bodies that donate are fantasised as entities that they may not be. Written essays, photos and medical histories form the basis of this imagination, for that is all clients have on which to base their decisions when choosing a donor. They fashion various metrics or go off of their limited knowledge in order to decide. Ultimately, they choose the "right" donor for them at that time. These donors can change as the trying-to-conceive process continues (donor specimens run out, not getting pregnant, a new donor piques interest).

Bodies are geographical sites and the geographies of bodies are essential in exploring reproduction. As such, they are complicated terrain to navigate. Donating bodies, receiving bodies, the exterior and interior of bodies all come into play when discussing artificial insemination. The body is imagined, medicalised and fragmented. A body of one can become a body of two. That's the aim of artificial insemination, after all.

Intimacy and sites of insemination

Making the choice of where to inseminate can be complicated for some, but for others, quite easy. Many SMBC and couples chose to inseminate at home when health issues were either not a factor or unknown. Remember that this cryobank was unusual in that it allows delivery of specimens to a private address and did not require medical supervision, an attractive feature for some participants.

It was getting to be the one who got my wife pregnant and vice versa. For us, it was more intimate at home—not in a sex way, just in a "not clinical" way—and there's the bonus of being able to stay right where you are when you're done and have a nice nap in your own cozy bed. Also, it's easier to do the inseminations whenever you determine that you need them versus whenever the clinic can fit you in. That was what drove us to DIY [do it yourself] in the first place. So, for us, it was less stressful to DIY in the comfort of our own home. And it cut down on the cost too since it cut out the cost of the doctor. Our clinic was even charging to thaw the vials—which, if you've ever DIY, you know is ridiculous.[10]

(Minnie, 35, LC)

Privacy, intimacy and cost were three major themes that came through regardless of whether the client chose to inseminate at home or with a reproductive endocrinologist. The participants (15 women) in this study were split on their sites of insemination. While some women chose to try at home, others used a reproductive endocrinologist. Some had previously tried at home without success and had moved to using a RE. Another went from employing a reproductive endocrinologist to trying at home:

> [A different cryobank] was so expensive and with all the doctor's fees, I had to take out a loan. So, I'm so happy that [the cryobank] will ship to my house. This will be my first time inseminating at home. I'm a little nervous about it but hopefully it will work this time. Going to the RE was so stressful! I always felt like the nurses were looking at me a certain way because I'm not married.
>
> (Brenda, 30, SMBC)

When asked about discrimination from reproductive endocrinologists or clinics, one participant responded that she had contacted clinics beforehand to ask if they are open to working with single mothers wanting to do donor insemination:

> Most are, even if they don't make it obvious in their advertising. Many do require a psychological "evaluation" for anyone using donor gametes, but from what I've heard it's usually just a psychologist asking you about your plans for how you will raise the child (i.e., whether you plan to tell the child from an early age that he/she is donor-conceived). I was fortunate and found a clinic where the psych consult was optional.
>
> (Mona, 41, SMBC)

Some participants in this study felt that the psychological evaluation was invasive and a deterrent to using a RE.

Geographies of insemination are important to consider when discussing artificial manipulation of reproduction. Where people choose to inseminate and why they choose to inseminate in that place can produce new geographies of reproduction. Instead of an intimate or private act, it can become very public in a clinical environment with a doctor/nurse and a catheter.

In the clinic, a speculum is placed in the vagina and a catheter threaded through the cervix. The doctor then injects the "washed" sperm into the catheter with a syringe, placing it in the uterus. The client is then to lie still for five to 15 minutes and then put their undergarments and lower body clothes back on. The lack of intimacy at the RE's office and clinical nature was outweighed by expertise for some. Others found it "cold" and the "wrong conditions to make a child".

For those who chose the reproductive endocrinologist route, medical testing was seen as a bonus and often a tradeoff for intimacy or privacy.

> We debated this for a long time before making a decision. I started seeing an RE and talking to them about their testing and to make sure everything was ok before we actually started the process made me feel better. Knowing ahead of time if something would prevent me from getting

pregnant made everything easier. Also, we had some friends that did it at home with a sperm donor and it took them six tries, six times close to $500 a time plus your ovulation kits and such adds up quickly. The insemination itself was $450 and we got pregnant the first try.[11] We also said if it didn't work at the doctor's, we would try at home. It's also more of a peace of mind thing for me knowing that I had an excellent team of doctors and nurses trying to help make our dreams come true, up until the day I was released from their care at nine weeks. I did not want to leave them!

(Kim, 26, LC)

When deciding whether to inseminate at home or use a RE, the choices are usually based on a combination of three things: intimacy, cost and efficiency. Those who choose to inseminate at home usually do so to have a more intimate environment for conception or to save costs. Those who employ a RE tend to do so in order to have the best chances for conception—cost is usually sacrificed in these cases.

Specimen mobilities

Travel was another component considered in determining sites of insemination, but additionally demonstrates mobility. Some clients traveled abroad to either access fertility drugs less expensively or without a prescription, while others skirted the legal restrictions in their home country:

I'm excited to be heading to Hawaii. After eight failed attempts and two miscarriages in New Zealand with the fertility clinic, I am running out of money so going to do an at home (at a hotel) ICI. I did this in Vegas a year ago with [donor]. This time I have picked [another donor]. New Zealand has a law that no bodily fluids can be shipped in or out so that means the one chain of clinics here can charge thousands. There is also a two-year waiting list for a sperm donor and you also only get to choose from four profiles.

(Emma, 32, SMBC)

After donation, the next instance for specimen mobility occurs once the donor has been selected and purchased. The vials are then shipped via a dewar containing frozen nitrogen to keep the specimen(s) frozen. The dewar containing the vials is then shipped to a residence or a clinic in one or two-day shipping via a commercial courier (e.g., Federal Express or United Parcel Service). For those who choose to inseminate at home, instructions are included in the box, which detail how to check and thaw the vial. The dewar is rented and guaranteed for a week. After that period of time expires, daily charges accrue ($20 per day) and the temperature of the sperm is not ensured. If the sperm does not arrive on time due to shipping difficulties, the shipping costs are returned. In the event that the dewar does not arrive due to cryo-bank error, the cost of the sperm and shipping is refunded, but the cycle and perhaps the donor (if no longer available/donating) is lost.

While clients are not allowed to pick up vials at the cryobank in order to preserve donor anonymity, clients may have to travel before insemination. Although located in the United States, the cryobank studied here does ship internationally for an additional cost. Some clients travel to places such as Mexico or Canada to obtain fertility drugs without a prescription or at a lower price. Those who use a reproductive endocrinologist may travel for hours in order to receive fertility treatment, ultrasounds or IUIs:

> Since we live hours from civilization, the closest RE is about a three hour drive.
>
> (Hannah, 35, HC)

Geographies of mobility are important. Artificial insemination cannot happen without circulation, transportation or motility. Mobilities are vital to under-standing how reproduction happens in an "artificial" environment. Without donation, without transportation, without insemination (via syringe, men-strual cup or catheter), sperm do not get to where clients desire them to go.

Conclusion

The trend of artificial insemination is increasing and more women and cou-ples are turning to AI in hopes of getting pregnant. Geographers need to understand how the spatialities of artificial insemination, including donation/reception of specimens, site of insemination and the associated mobile aspects affect bodies and the spaces (whether the home or the clinic) and processes associated with conception.

In artificial insemination, clients are able to select donors with desired traits (e.g., one participant stated that she chose her donor because she wanted a "dark haired, dark eyed baby"). While this may be eugenics in some form or simply consumer preference, donor selection shows how the body can be seen as a compilation of characteristics and broken down into superficial compo-nents. Steve Pile (1996, 146) argues the "contours of the body are the contours of society". In this case, the desired donor body demonstrates the desires of society. The imagined donor is part of the fantasised child.

Geographers work to disrupt the naturalisation of the body. Geographies of the bodies that receive sperm show how the "natural" process of conception is manipulated by many seeking pregnancy. The sometimes intense and con-suming preparation of the body in anticipation of pregnancy destabilises the notion that conception is easy and a simple process. Pregnancy can be a very calculated situation.

Following Tine Tjørnhøj-Thomsen (2005), Svendsen (2007, 38) notes that donation transgresses "boundaries between the public, the private and the intimate". This disruption is also seen in sites of insemination. The site of home for insemination is often seen as intimate and private and, as such, touted by some research participants as a better way to bring "a child into the world". The choice of a clinic can be contradictorily sterile, intimate and public all at the same time. While maybe not a private space, a clinic can increase odds of conception and be a more desired site of insemination.

Artificial insemination demonstrates the mobility of bodies as well as of body parts. The fragmentation of bodies is important to the mobilities involved in "artificial" forms of reproduction. The ways that movement occurs include donation, delivery, potential travel to site of insemination and then insemination itself. Mobility, in its different forms, shows how artificial insemination is predicated on changing ways and understandings of reproduction.

Those of us who study the body and other intimate geographies need to recognise the importance of artificial insemination. Manipulation of reproduction is becoming increasingly common. This phenomenon is becoming more prevalent and medical advances over the past few years (including the recent uterine transplants in Sweden and the U.S.) are changing reproduction and changing understandings of reproduction. As such, the spatialities of reproduction are in need of more geographical examination.

Notes

1 Cytomegalovirus (CMV) is a member of the herpes virus family that includes cold sores, chicken pox and infectious mononucleosis. CMV causes more birth defects and congenital disabilities than many other well-known diseases (e.g., hearing loss, neurological abnormalities or decreased motor skills). To reduce the risk of adverse pregnancy outcomes, a CMV negative client should use a CMV negative donor, while a CMV positive patient can use either.
2 A LH (luteinizing hormone) surge means ovulation should occur soon. This timing is often dependent on age.
3 Doctors time insemination differently than at home clients because they can more precisely control the release of the egg with the trigger shot.
4 Progressive motility is defined as forward movement or movement in large circles.
5 In open donation, cryobanks release the identity of the sperm donor (with their permission, which can be changed at any time) to a donor-conceived child once the child reaches age 18. Known donors are personally selected by the client but do not have sexual intercourse with the sperm recipient.
6 Anonymity may be considered as a possible perk because parentage will not have to be disclosed at some point nor is there pressure to reveal siblings (unless a sibling registry is employed) or donor.
7 This research was conducted before this cryobank was sold to one of the largest cryobanks in the United States, which caused prices to increase dramatically much to the dismay of customers.
8 Pseudonyms are used.
9 The extended profile can be purchased for $50, if available. Basic profiles are free.

10 At-home instructions state that once delivered, one is immediately to open the dewar (filled with frozen nitrogen) to ensure that the proper donor vial was sent and then leave untouched and upright until time of insemination. Fifteen minutes prior to insemination, the vial should be removed from the tank and thawed. The first few minutes of the thaw should be on a flat surface until the vial can be brought to body temperature in the hands. Insertion into the vagina is via syringe or soft cup device.

11 Intrauterine insemination has a higher chance of resulting in pregnancy than intracervical insemination.

References

Almeling, Rene. 2011. *Sex Cells: The Medical Market for Eggs and Sperm.* Berkeley: University of California Press.

American Pregnancy Association. 2017. "Emotions of Infertility." Accessed 24 August. http://americanpregnancy.org/infertility/emotional-faq/.

Berg, Barbara J. 1995. "Listening to the Voices of the Infertile." *Reproduction, Ethics, and the Law: Feminist Perspectives,* edited by Joan C. Callahan, 80–108.

Bergmann, Sven, 2011. "Fertility Tourism: Circumventive Routes that Enable Access to Reproductive Technologies and Substances." *Signs: Journal of Women in Culture and Society,* 36(2): 280–289.

Blake, Lucy, Martin Richards, and Susan Golombok. 2014. "The Families of Assisted Reproduction and Adoption." In *Family-Making: Contemporary Ethical Challenges,* edited by Francis Baylis and Carolyn McLeod, 64–85. Oxford: Oxford University Press.

Boyer, Kate. 2010. "Of Care and Commodities: Breast Milk and the New Politics of Mobile Biosubstances." *Progress in Human Geography* 34(1): 5–20.

Chadwick, Ruth and Alan O'Connor. 2015. "Biobanking across Borders." In *Bodies Across Borders: The Global Circulation of Body Parts, Medical Tourists and Professionals,* edited by Bronwyn Parry, Beth Greenhough, and Isabel Dyck, 15–28. London: Routledge.

Colls, Rachel. 2006. "Outsize/outside: Bodily Bignesses and the Emotional Experiences of British Women Shopping for Clothes." *Gender, Place & Culture,* 13(5): 529–545.

Colls, Rachel and Maria Fannin. 2013. "Placental Surfaces and the Geographies of Bodily Interiors." *Environment and Planning A* 45(5): 1087–1104.

Cresswell, Tim. 2006. *On the Move: The Politics of Mobility in the Modern West.* London: Routledge.

Cresswell, Tim, and Peter Merriman, eds. 2011. *Geographies of Mobilities: Practices, Spaces, Subjects.* London: Ashgate Publishing.

Culley, Lorraine, Nicky Hudson, and Floor Van Rooij. 2009. "Introduction: Ethnicity, Infertility, and Assisted Reproductive Technologies." In *Marginalized Reproduction: Ethnicity, Infertility and Reproductive Technologies,* edited by Lorraine Culley, Nicky Hudson, and Floor Van Roojj, 1–14. London and Sterling, VA: Earthscan.

D'Arcy, Janice 2011. "Redhead Sperm Donors Not Wanted at World's Largest Sperm Bank" *The Washington Post.* https://www.washingtonpost.com/blogs/on-parenting/post/redhead-sperm-need-not-apply/2011/09/19/gIQAqXCSiK_blog.html?utm_term=.77d1c07d8c7a

Daniels, Cynthia R., and Janet Golden. 2004. "Procreative Compounds: Popular Eugenics, Artificial Insemination and the Rise of the American Sperm Banking Industry." *Journal of Social History* 38(1): 5–27.

Douglas, Mary. 1966. *Purity and Danger.* London: Ark Books.

Fannin, Maria. 2013. "The burden of choosing wisely: Biopolitics at the Beginning of Life." *Gender, Place & Culture* 20(3): 273–289.

Foucault, Michel. 1973. *The Birth of the Clinic: An Archaeology of the Human Sciences.* New York: Vintage.

Inhorn, Marcia C. and Pasquale Patrizio. 2009. "Rethinking Reproductive 'Tourism' as Reproductive 'Exile'". *Fertility and Sterility* 92(3): 904–906.

Jungheim, Emily S., Man Yee (Mallory) Leung, George A. Macones, Randall R. Odem, Lisa M.Pollack and Barton H. Hamilton. 2017. "In Vitro Fertilization Insurance Coverage and Chances of a Live Birth." *JAMA* 317(12): 1273 doi:10.1001/jama.2017.0727

Longhurst, Robyn. 1997. "(Dis)embodied Geographies." *Progress in Human Geography* 21(4): 486–501.

Longhurst, Robyn. 2004. *Bodies: Exploring Fluid Boundaries.* London: Routledge.

Longhurst, Robyn. 2012. *Maternities: Gender, Bodies and Space.* London: Routledge.

Longhurst, Robyn, and Lynda Johnston. 2014. "Bodies, Gender, Place and Culture: 21 Years on." *Gender, Place & Culture* 21(3): 267–278.

Martin, Emily. 2001. *The Woman in the Body: A Cultural Analysis of Reproduction.* Boston: Beacon Press.

McDowell, Linda. 1999. *Gender, Identity and Place: Understanding Feminist Geographies.* Minneapolis: University of Minnesota Press.

Parry, Bronwyn. 2015. "A Bull Market? Devices of Qualification and Singularisation in the International Marketing of US Sperm." In *Bodies Across Borders: The Global Circulation of Body Parts, Medical Tourists and Professionals,* edited by Bronwyn Parry, Beth Greenhough, and Isabel Dyck, 53–72. London: Routledge.

Pile, Steve. 1996. *The Body and the City: Psychoanalysis, Space, and Subjectivity.* London: Routledge.

Philo, Chris. 2000. "The Birth of the Clinic: An Unknown Work of Medical Geography." *Area* 32(1): 11–19.

Rose, Gillian. 1993. *Feminism and Geography: The Limits of Geographical Knowledge.* Minneapolis: University of Minnesota Press.

Svendsen, Mette N. 2007. "Between Reproductive and Regenerative Medicine: Practising Embryo Donation and Civil Responsibility in Denmark." *Body & Society* 12(4): 21–45.

Tjørnhøj-Thomsen, Tine. 2005. "Close Encounters with Infertility and Procreative Technology", in *Managing Uncertainty—Ethnographic Studies of Illness, Risk and the Struggle for Control,* edited by Vibeke Steffen, Richard Jenkins and Hanne Jessen, 71–91. Copenhagen: Museum Tusculanum Press, University of Copenhagen.

Tober, Diane M. 2001. "Semen as Gift, Semen as Goods: Reproductive Workers and the Market in Altruism." *Body & Society* 7(2–3): 137–160.

Tourielle, Claire. 2017. "A Sperm Bank Wants More Redhead Donors to Save Gingers from 'Extinction'" *International Business Times.* https://www.ibtimes.co.uk/sperm-bank-wants-more-redhead-donors-save-gingers-extinction-1630975

Urry, John. 2007. *Mobilities.* Cambridge: Polity.

Valentine, G., 2008. "The Ties that Bind: Towards Geographies of Intimacy". *Geography Compass* 2(6): 2097–2110.

Woliver, Laura R. 2002. *The Political Geographies of Pregnancy.* Champaign: University of Illinois Press.

World Health Organization. 2010. *WHO Laboratory Manual for the Examination and Processing of Human Semen.* Fifth ed. Geneva, Switzerland: WHO Press.

4 Behind closed doors

The hidden needs of perimenopausal women in Ghana

Amita Bhakta, Brian Reed and Julie Fisher

Introduction

Women going through the perimenopause are unnoticed by society, but deserve attention in geographies of reproduction. The perimenopause includes "the period immediately before the menopause (when the endocrinological, biological and clinical features of approaching menopause commence) [up until the] first year after menopause" (Utian 1999, 284). The menopause marks the permanent cessation of menstruation, when a woman has not menstruated for 12 months. In reproductive geographies, social aspects of reproduction are the focus of most of the literature.

The perimenopause is an individual experience that is not shared with others, nor is it visible to geographers (DeLeyser and Shaw 2013). Women's experiences of the perimenopause are experienced in isolation as though each woman is in a "bubble" on her own. Feminist oral history methodologies, adopted in this research, seek to burst these isolated bubbles and validate women's individual perceptions and experiences of the perimenopause.

Drawing upon narratives of women, this chapter is about the water, sanitation and hygiene (WASH) needs of women during the perimenopause. Based upon doctoral research conducted in Accra and Kumasi, Ghana, this chapter looks at the everyday hidden geographies of the perimenopause through women's use of water and sanitation during this particular reproductive stage. Using a feminist approach, this research employed oral history interviews, participatory photography and participatory mapping, complemented with data from observations. The research objectives were focussed around three main questions: 1) What are the hygiene needs of perimenopausal women? 2) What are the water and sanitation needs of perimenopausal women? 3) How are these needs of perimenopausal women affected by human, natural, financial, social and physical factors? This chapter goes further to ask, what are the gendered and spatial aspects of the perimenopause as demonstrated through WASH?

Gender-water geographies are well-discussed, as illustrated in the special issue of *Gender, Place and Culture*. In this issue, Harris (2009) calls for gender and water theorists to pay greater attention to the effects of neoliberalisation in water governance. Sultana (2009) draws upon fieldwork on arsenic in water to explore how types of water and ideological constructs of masculine and feminine affect people's relation to water, and how ecological and social factors interact to reinforce inequity in relation to gender and water. Alhers and Zwarteveen (2009) draw upon agricultural water use to challenge the neoliberalisation of water through a feminist perspective. This chapter expands these debates to explore gender–water relations, specifically exploring women's relation to water through hygiene during the perimenopause. By examining how limited water and sanitation infrastructure in patriarchal Ghana dictates perimenopausal women's movements through space, this chapter builds upon discussion by Laws (1993) who explored how patriarchal social structures limit women's spatial mobility.

Over the past decade, there has been an increasing focus on menstrual hygiene management (MHM) in the WASH sector. This work has predominantly focused on the hygiene needs of adolescent schoolgirls experiencing their first menstrual period or menarche, marking the start of the reproductive life stage (for instance, Sommer 2011; Crofts and Fisher 2012; Kidney et al. 2013; Jewitt and Ryley 2014). Perimenopausal women, who are aged between 47 and 51 on average (Loue and Sajatovic 2008), account for an estimated 3.6 per cent (UN 2015) of the 2.4 billion people who lack access to improved sanitation and the 663 million people without improved drinking water (UNICEF and WHO 2015). However, age is only an approximate indicator of perimenopause, as the age at which symptoms are experienced by women can vary.

This chapter explores the hidden geographies of perimenopausal women as demonstrated through WASH use in Ghana. The country has made progress in the provision of an improved water supply to over 90 per cent of the population by 2015 but sanitation provision lags behind, with less than 60 per cent of Ghanaians having access to safe management of excreta and solid and liquid waste (WHO/UNICEF 2017). The level of "adequate" provision means that basic needs might be met, but any additional needs of perimenopausal women may still not be met.

Women in Africa are conventionally associated with particular spaces such as the kitchen (Robson 2006) or water collection points (Fisher 2008). Women are not typically associated with latrines and bathrooms, and their WASH needs—particularly during the perimenopause—are spatially, temporally and physically hidden. Hygiene-related activities such as toileting or bathing are conducted secretly and in private. Knowledge about practices such as menstrual hygiene materials used are not shared amongst women due to taboos around the menopause and menstruation. Existing discourse on gender mainstreaming in sanitation provision is consistent with a neglect of women's needs across the life course, and especially the perimenopause. Hygiene is vital to the everyday geographies of the perimenopause as performed within different sanitation spaces, regardless of the degree of sanitation provision made.

The Perimenopause

Biological understandings

The perimenopause is understood biologically through physical changes in the body, marking the transition from the reproductive to the non-reproductive stage (Utian 1999). The perimenopause is often referred to as "the change of life" or the "Change" as the female body ages and begins to lose its fertility (Greer 1991; Torpy et al. 2003). Perimenopausal women experience a wide range of physical symptoms, including hot flushes (Stearns et al. 2002), day and night sweats (Prior 2005), urinary incontinence (Sampselle et al. 2002), heavy bleeding during periods (Duckitt 2010), irregular periods (Torpy et al. 2003), vaginal dryness and dry skin (Avis et al. 2009).

Cultural understandings of the perimenopause

Some reproductive geographies are explicitly evident and visible to others. Pregnancy, a visible stage of reproduction, is discussed in this book through choice of birth centres (Hazen) and homebirths (Whitson) or the lived experiences of pregnant graduate students at university (Merkle). Unlike pregnancy, the majority of symptoms of perimenopause are not "visible" to the outsider, and are therefore not commonly discussed. Women have to self-identify as perimenopausal, as there is ambiguity over these symptoms and variation in the age at which they are experienced. Women may be reluctant to accept the perimenopause as it marks the end of reproduction and loss of the ability to have children. Acknowledgement that a woman is perimenopausal is not guaranteed, and they may not overtly identify as such, particularly in the early stages when they are still menstruating, albeit irregularly. Not only are symptoms not obvious, they may be actively hidden.

The perimenopause in the Global North

The "hiddenness" of the perimenopause as a reproductive stage of life is discussed in literature on cultural perspectives from the Global North. When women go through the "Change" (Greer 1991), the end of reproductive years is marked privately, without any public rite of passage. Often it is difficult for a woman to talk to other women about her experiences, and so women have to deal with their changes alone, "behind closed doors" and secretively. Whilst Greer writes as a feminist from the Global North, the limited knowledge about women's perimenopausal experiences in the Global South suggests that women also find it challenging to discuss the perimenopause in low-income countries.

Scholars have provided their own perspectives on what the menopause means. For Greer (1991), the menopause can be seen to mark the death of the womb and the ovaries, thus, part of a woman is believed to die. The perimenopause is seen as a time for stock-taking and a time to be fully aware of what is going on, confronting the problem of ageing (Greer 1991). The menopause is often characterised by feelings of loss: of menstruation, child-bearing, youth and in turn, womanliness (Mackie 1997). Thus, menopausal women often find themselves ignored by their partners as sexual beings; the loss of their femininity leaves them invisible (Greer 1991). Invisibility of older women in film and fiction, which often give preference to younger women, contributes to silences surrounding menopause in wider society (Greer 1991; Rogers 1997). As DeLeyser and Shaw (2013, 504) summarise, the "meno-pause is stigmatised for those who experience it first hand: women". These factors collectively contribute to the limited exploration of the menopause as a reproductive geography (DeLeyser and Shaw 2013). The limited geographical discourse stipulates that ageing bodies are coded according to what they can or cannot do (Laws 1995). Jones et al. (1997) state that the menopause is a marker of when women are "too old" to become mothers. This literature does not go beyond the internal, individual bodily experiences of the menopause and does not relate the menopause to society, space, place and the environment.

The perimenopause in Africa

Cultural beliefs surrounding the perimenopause and the menopause vary sig-nificantly between different African countries, with the menopause viewed both positively and negatively according to different cultures. Africans con-sider the menopause itself, including the cessation of periods and symptoms such as sweats and flushes as part of the ageing process for women (Wambua 1997). In Nigeria, Hausa women see the menopause as a marker of freedom, for they are no longer restricted in Hausa culture to remain within the boundaries of the homestead, as they have been from the day of marriage. Yet, the menopause also signals that a woman is no longer recognised as sexually active, marked by the prohibition of coitus from this point onwards (Johnson 1982). For the Wikidum women of Cameroon, the menopause brings sexual freedom, providing the opportunity for extra-marital relations without the risk of pregnancy. In Cameroon, tribes such as the Bamilikes see the menopausal woman as being as wise as a man and hence her position rises to that of a leader in society but for the Diis women of the Adamoua plateau, the inability to bear children is likened to being a man and meno-pausal women attract sympathy (Wambua 1997). Menopause itself is asso-ciated with ill health in some cultures, which argue that menstruation is part of the cleansing process that keeps women healthy (Wambua 1997).

Understandings of the menopause: The Ghanaian perspective

Issues that relate primarily to women are not widely talked about in Gha-naian society. Attitudes and beliefs about matters relating to menstruation and the menopause mean that these topics are taboo and not publicly discussed. The silences which surround menstruation are demonstrated through the limitations placed upon women: being forbidden from entering a house, cooking, associating with men and participating in social activities, as menstruating women are considered unclean (Bhakta et al. 2016).

The term "elder" is a mark of respect. In Ghana, women are considered to be "elders" by the age of 50–52, around the time of menopause. Women during perimenopause can be considered "elderly", although according to local customs, a person becomes elderly once they are the oldest family member, regardless of their age (Lartey, personal communication).

Records of Ghanaian beliefs in the past about the menopause are found in a few studies dating from over 50 years ago. The cessation of menstruation marked a new stage in a woman's life (Field 1960); women are now free to reside with men, participate in rituals from which they were prohibited whilst menstruating and visit ancestral shrines (Nukunya 1969). The loss of fertility denoted women as being asexual, unable to fulfil what is regarded as their primary task: childbearing (Field 1960). This was particularly poignant in a patriarchal setting where men had a desire for many children, and the cessation of menstrual periods could initially be mistaken for pregnancy (Belcher 1974). Recent studies link hot flushes, night sweats, insomnia, anxiety, irritability, mood swings and short-term memory problems to witchcraft (Adinkrah 2017). Prevalence of these beliefs promotes a culture of silence about the menopause (Bhakta et al. 2016).

Women and water

WASH is not just the provision of infrastructure; current debates focus also on equity and inclusion within service provision. Equity recognises how individuals differ from each other and identifies the support and resources required for them to realise their rights to WASH. Inclusion seeks to ensure that all, including the disadvantaged, discriminated against and marginalised, have access to services and that they have an active engagement in wider processes relating to these rights (Gosling 2010).

Women's experiences of the perimenopause, as well as other life stages, can be revealed through the study of the physical, spatial, and temporal elements of WASH use. To ensure equality for all women at reproductive health stages in a WASH context, the needs of perimenopausal women have to be assessed, in keeping with the SDG agenda to "leave no-one behind" (WSSCC and FANSA 2015; UN 2016b).

Women going through the perimenopause will have different experiences of their built environment compared with women of other age groups and with men. They have specific needs in relation to bathing, laundry, (menstrual) hygiene, solid waste management and sanitation, which become acute when these basic services are not readily available, especially at night. However, these women may be uncomfortable discussing these needs as the perimenopause is generally socially taboo. As the symptoms of perimenopause are not "visible" to the outsider, perimenopausal women may be overlooked. Women may also be unwilling to share their views with the (often male) providers of these services.

Hidden knowledge

Defining hidden knowledge: the Johari Window

"Hidden knowledge" is that which is known or held by particular individuals or groups, but is not shared with other people or recorded in writing. Hidden knowledge can be modelled by a simple matrix, the Johari Window (Figure 4.1) (Shenton 2007) that consists of four panes for the forms of knowledge: "public", "blind", "hidden" and "unknown" (Luft 1961):

(1) "Public" or "open" knowledge is known by yourself and by others. Whether a water supply system is working or not is public knowledge, obvious to all.
(2) "Blind" knowledge is known to others but not to yourself. A water engineer may be aware of the benefits of sanitation, but the community may not know these benefits.

	Known to self	Unknown to self
Known to others	Public, or open knowledge	Blind knowledge
Unknown to others	**Hidden knowledge**	Unknown knowledge

Figure 4.1 The Johari Window

(3) "Hidden" knowledge is known by yourself, but not revealed to others. The WASH needs of perimenopausal women are hidden knowledge because women who are perimenopausal or menopausal are aware of their own personal issues, but others (including other perimenopausal women) are not.
(4) "Unknown" knowledge is not known by yourself or others. The best way to dispose of used sanitary pads in a low-income context is unknown to the women using them. The WASH professionals also do not know the best methods and may not even be aware that appropriate disposal is an issue.

WASH needs of perimenopausal women as hidden knowledge

The perimenopause marks the end of women's fertility and is experienced individually by women alone (Greer 1991). Cultural taboos and private hygiene practices mean that knowledge of the perimenopause is physically hidden in space. Further, unlike puberty or pregnancy, there are usually no visible, physical markers on the body of the perimenopause which enable others to identify it, nor does the perimenopause leave a trail in public space. Perimenopausal experiences are perhaps more akin to the "geology" than the "geography" of reproduction and require a bit more digging, relying on the elements that rise above the surface to indicate what is happening underneath. WASH practices during the perimenopause, such as additional bathing or laundry, are undertaken by women, but each may perform them individually, in private, making these practices "hidden knowledge", in terms of both actual service provision and in the literature. Women seldom speak about their perimenopausal experiences with each other or with outsiders (DeLeyser and Shaw 2013), limiting the social understanding of the perimenopause in wider society. Menstrual hygiene management for adolescent girls has received greater attention relating to WASH than the experiences of perimenopausal women. Exploring these experiences requires perimenopausal women's bubbles to be burst and this requires the effective participation of those who experience the perimenopause.

In the absence of explicit human interaction during the perimenopause (unlike conception, pregnancy, birth and breastfeeding), experiences of the perimenopause can be explored indirectly through women's interactions with the built environment, focussing upon how hygiene needs arising due to the perimenopause are met (or not met) through WASH infrastructure use. Exploring the WASH needs of perimenopausal women reveals women's individual knowledges about the built and natural environment and reflects their particular interpretation of their surroundings (Braun 2002).

Academic perspective: A lack of literature

A systematic literature search in 2013 produced little literature linking WASH and the perimenopause. Fifty-nine search term combinations included: "menopause", "perimenopause", "older women", "ageing", "water supply", "water and sanitation", "hot flushes", "menstruation" and "bleeding". Various combinations of these terms, including those of relevance to the perimenopause were used to search specialist WASH collections such as the Water, Engineering and Development Centre (WEDC) Knowledge Base and IRC WASH (see Table 4.1). Both collections relate to WASH, such as water supply, MHM, solid waste and faecal sludge management. Conventional databases were also searched: Science Direct, Zetoc, Google Scholar, Medline, ProQuest, Web of Knowledge and Geobase and Compendex (see Table 4.2).

The limited literature on low-income countries focuses on perimenopausal symptoms, including, for example, India (Bairy et al. 2009), Pakistan (Wasti et al. 1993), Bolivia (Castelo-Branco et al. 2005), Thailand (Chompootweep et al. 1993), Brazil (da Silva and d'Andretta Tanaka 2013), Egypt (Gadalla 1986), and Zimbabwe (Moore 1981). The literature focussing on Africa indicates cultural perceptions (Wambua 1997) that may reinforce the hiddenness of WASH experiences of perimenopausal women. More common symptoms of the perimenopause, such as hot flushes and night sweats, are recognised widely as signs of natural ageing, but not as markers of the perimenopause (Wambua 1997). The lack of any significant body of literature on this topic means the needs of perimenopausal women were/are still hidden.

Table 4.1 Sample results of literature search from specialist WASH collections

Search terms	WEDC Knowledge Base	IRC WASH Database
Menstrual hygiene management	25 retrieved 25 relevant	5526 retrieved 34 relevant
Perimenopause AND change of life	0 retrieved 0 relevant	0 retrieved 0 relevant
Ageing AND bleeding AND menstruation	0 retrieved 0 relevant	23 retrieved 0 relevant
Ageing AND perimenopause	0 retrieved 0 relevant	927 retrieved 0 relevant

Table 4.2 Sample results of literature search of non-WASH specific databases

Search term	Science Direct	Zetoc
Menstrual hygiene management	1817 retrieved 12 relevant	22 retrieved 18 relevant
Perimenopause AND change of life AND sanitation	9 retrieved 0 relevant	0 retrieved 0 relevant
Ageing AND sanitation AND bleeding AND menstruation AND developing countries	18 retrieved 0 relevant	0 retrieved 0 relevant
Ageing AND perimenopause AND developing countries AND water supply	31 retrieved 2 relevant	0 retrieved 0 relevant

Methodology and case study

Overall, the literature on the menopause in Ghana focuses predominantly on medical and cultural understandings rather than WASH experiences. More in-depth study requires participative methods adapted to cater for the taboos and sensitivities around the perimenopause and menstruation in Ghana. Chambers (1997) calls for development professionals to "put the first last", and to empower individuals who are weak. Participatory research explicitly focuses on working with marginalised individuals who may be "vulnerable" (Kindon et al. 2007). In this case, women may not appear to be or see themselves as "weak", "marginal" or "vulnerable" but the topic (if not the individuals) is socially excluded. In seeking to understand the needs of perimenopausal women through WASH, this study followed a participatory approach, applying "a democratic commitment to break the monopoly on who holds knowledge and for whom social research should be undertaken" (Fine 2008, 215).

Feminist research

Feminist research principles informed the methodology; research should be *with* women rather than *on* women.

> Feminist research is about the development and construction of knowledge founded upon the relationship between women's everyday experience, academic knowledge, political power and social action. This methodological approach facilitates the *central involvement of the women, who are active participants in the social construction of knowledge, empowerment and social change.*
>
> (O'Neill 1996, 131, authors' emphasis)

Hidden knowledge as a concept is not discussed in the realm of geography. Examining the needs of perimenopausal women explores hidden knowledge that is not recorded in the literature. The geographies of reproduction can note lessons from researching the needs of perimenopausal women. Feminist methods, which are participative and place the research focus on the experiences of women (Stanley and Wise 1983), can effectively reveal women's hidden reproductive geographies. Oral history narratives of WASH use give voices to perimenopausal women's intimate experiences of their environment. Participatory mapping provides a spatial analysis of perimenopausal women's issues with community-based infrastructure and services, and perimenopausal women taking photographs brings a visual element to a hidden experience. Collectively, these methods enable built environment professionals to gain an understanding of perimenopausal women's infrastructural needs, which is needed to improve service provision.

Case study and data collection

Data was collected in two low-income communities: La, Accra (an urban setting), and Kotei, Kumasi (a peri-urban setting). La and Kotei both have limited WASH services. La lies within the La Dade-Kotopon Municipal Assembly (LaDMA) district of Accra, on the coast. Ghana's 2010 Census recorded that 44 per cent of the broader LaDMA district population used public toilets, 24 per cent drank sachet water (the local equivalent of bottled water in small sealed plastic bags), 31 per cent had piped drinking water and 9 per cent collected drinking water from a standpipe (Ghana Statistical Service 2014). La has the highest population density in LaDMA with 120 people per square kilometre. LaDMA was estimated to have a population of 333,817 in 2013. Infrastructural services are overstretched in high-density compound housing in the low-income township of La (LaDMA 2013). Water sources originally consisted of springs, but now tend to be a piped metered community-based connection, which is billed each month. Seventy per cent of those interviewed in La use one of two shared toilet blocks: one has nine toilets and no doors, and the second is the largest with 48 toilets, able to serve almost the whole of La.

Kotei is an ex-rural community which has become part of the city of Kumasi in the Ashanti region. With an old and a new town, Kotei sits adjacent to the Kwame Nkumrah University of Science and Technology campus (Awuah et al. 2014). Almost half (47 per cent) of residents in the main parts of Kotei rely upon three public toilets; 35 per cent of household latrines are WCs and 18 per cent are pit latrines (Leathes 2012). The main river source was the river Daakye, which still flows but is not used today. Kotei residents still use some springs, which were the main source of water prior to the introduction of a pipe-borne water supply in 1978 (Awuah et al. 2014), along with a communal water storage tank and a rainwater harvesting system. Despite this, it was observed that the predominant drinking water for both settings was sachet water.

The women in this study were either unemployed and looking for work, retired or self-employed in low-income, marginal work such as petty trading. Across the two communities, 14 oral history interviews, four participatory mapping sessions and five participatory photography exercises were conducted, exploring different aspects of women's needs. Oral history interviews identified the intimate hygiene practices undertaken to cope with perimenopause, the daily patterns and (ir)regularities of these hygiene practices, and the more public issues faced with WASH infrastructure. The narratives were complemented with visual data from the photography and spatial data from the mapping sessions, which brought to light women's issues with community-based WASH infrastructure.

Hidden needs of perimenopausal women: What goes on behind closed doors?

Hygiene is the "practice of keeping oneself and one's surroundings clean, especially in order to prevent illness or the spread of disease" (Boot and

Cairncross 1993, 6). It stems from the Greek term *hygienios,* which refers to health or being healthful (Reed and Bevan 2014). The hygiene practices of perimenopausal women typically occur in household bathrooms where available, compound-based bath houses (several homes are built around a common courtyard or compound) and, predominantly, in public latrines. In Ghana, these hygiene practices are affected by inadequate access to WASH services. The studies showed women have three particular hidden hygiene needs that have been present throughout their adult life, but change dramatically as they enter the perimenopause: menstrual hygiene management, bathing and laundry. This section presents some preliminary findings on this, with scope for deeper exploration in the future into the geographies of the perimenopause, through WASH and beyond.

Menstrual hygiene management

The inclusion of MHM under WASH facilities is recent (House et al. 2014). Perhaps due to its highly gendered and private nature, this aspect of sanitary behaviour has not been included with other aspects of sanitation, such as handwashing and excreta disposal. Irregularities in the menstrual cycle and heavy bleeding in menstrual periods for perimenopausal women affects their MHM needs and techniques, medical intervention and access to infrastructure.

MHM techniques

Heavy menstrual periods, marking the entry to perimenopause, require MHM techniques that are different to those used during the reproductive years. In Ghana, access to commercial menstrual hygiene products (e.g., sanitary pads and tampons) is mainly restricted to women with high incomes. Less wealthy women rely on traditional techniques (e.g., rag cloths, cotton wool) perhaps combined with cheaper commercial products (often low absorbency sanitary pads), if they can afford them and they are available, to manage heavy menstrual flow. One older woman, who went through the perimenopause at a time when commercial sanitary products were not widely available, explained her use of cotton wool and cloth combined:

> I put the cotton on the cloth in order not to soil the cloth and when I'm going to take my bath I dispose of just the cotton whiles the cloth remains unsoiled.
>
> (Audrey[1], La, aged 70)

In some cases, women resorted to using very high absorbency products such as diapers:

It started flowing very heavily, and that was not how it was when it started, when it gets full; we were using diapers at the time and I change,
then it full again until the fifth day when it will reduce.

(Audrey, La, aged 70)

Fears of leaking and staining from techniques used prior to the perimenopause
led to the use of a combination of traditional and commercial materials:

Yes it [bleeding] changed a lot, using the pad wasn't even enough, I had
to support with pieces of cloth, and it flows in big clots.... I had it do
that to help soak up the blood.... I use both because of heavy bleeding,
else the pad alone would have been okay.

(Abla, Kotei, aged 50)

Heavy blood flow during the perimenopause required medical intervention in
some circumstances, with women going to hospital for treatment. Appropriate
commercial pads were bought for one woman by her daughter in an emergency, after arrival at the hospital due to a particularly heavy period:

What I used from the house was so much soiled up so we threw it away
then afterwards I started using the pad.

(Bertha, La, aged 60)

Access to infrastructure

Solid waste disposal

Menstrual hygiene management during the perimenopause raised concerns about
the disposal of menstrual hygiene materials. Multiple family-occupied compounds
are the predominant housing style in La and Kotei, which is typical of many
communities in Ghana (Gough and Yankson 2011). Lack of space in which to
construct household toilets within compound housing necessitated the use of
public, shared, pay-per-use community toilets for MHM. Public toilets lack
incinerators or other means to discreetly dispose of the secret waste of heavily
soiled cloths and pads, which women were embarrassed about others seeing:

I keep it on me till I get back (home) so that I can properly wrap it in a
poly bag and dispose it.

(Abla, Kotei, aged 50)

Solid waste management services at a community level were inadequate and,
at times, lacking in both Kotei and La. Most women in Kotei used a community dumpsite, so they carried their stained menstrual materials to it.
Waste collection services were provided at a household level in La, so women

could dispose of menstrual materials at home. Women raised concerns about the affordability of waste collection, which came into sharper focus when hiding heavily soiled materials during the perimenopause.

Access to latrines

Latrine design is important for effective MHM. Heavy menstrual flow led women to use the latrines with extra care:

> Any time after visiting the toilet, I end up staining the slab or the floor with blood and I usually feel bad thinking about another person coming to see that.... I try to cover the blood stains with something, either a tissue, or wash it away with water...
>
> (Abla, Kotei, aged 50)

Bathing

Infrastructure and bathing during perimenopause

The WASH sector is largely oblivious of the bathing needs of individuals (both men and women), with sparse discussion about suitable bathing infrastructure (Bhakta et al. 2017), despite the "WASH" acronym suggesting "washing" is the focus. Instead the majority of discussions in WASH revolve around water supply and sanitation infrastructure provision. Discussions on hygiene practice have largely focussed upon topics such as handwashing, MHM, hygiene promotion, faecal sludge management and drinking water supply. Bathing has been explored from socio-cultural perspectives in the Global North (Pickerill 2015), but the WASH sector has not explored individual bathing needs in low-income countries to enable effective infrastructure provision (Bhakta et al. 2017). Bathing is particularly important during the perimenopause. Women bathe more frequently than they typically would prior to entering the perimenopause, increasing their demand for water. Unpredictable symptoms, which can occur at any time (heavy blood flow, hot flushes, urine leaks and sweating) are often dealt with through bathing.

> When I am menstruating, I bath three times a day.
>
> (Abla, Kotei, aged 50)

Bathing during the perimenopause highlighted infrastructural issues faced by women. Women experiencing heavy flow during periods had particular need of appropriate infrastructure in order to bathe. Travelling beyond the local community was difficult for these women as a result:

> I had no issue (at home) because my bathroom is just here, but I get worried when travelling and menstruating. I do think about access to bathing area outside home.
>
> (Abla, Kotei, aged 50)

The perimenopause therefore restricted women's movements due to dependency upon bathing infrastructure. Women preferred to stay at home where they were certain of bathing facilities. The bathhouse within the home was the best available solution to privately wash away blood, keeping it hidden from view. Despite access to bathing facilities in the home, lack of supportive infrastructure through piped water supply and drainage made bathing a physical challenge. Poor design or absence of infrastructure presented social challenges for women as experiences that they wished to remain hidden in private would be exposed in public spaces against their will, in a setting where menstruation is taboo. The bathhouses used by some women in La and Kotei were in some cases disconnected from any form of drainage, with wastewater disposal in the open street of a densely populated community. Discussions revealed that this was particularly embarrassing for women who were experiencing heavy menstrual flow, as they did not wish others to see wastewater in the street, stained red with blood, and splashed upon their feet.

Perimenopausal symptoms also required women to bathe at irregular times:

> I sweat at night sometimes and when that happens I clean myself with a wet towel and, at certain time, I enter the bathhouse to pour water on myself.
>
> (Mansa, Kotei, aged 57)

This was difficult due to the irregular water supply. Constant water supply is compromised as a result of Ghana's power supply crisis, colloquially known as "dumsor", or on and off (Bayor and Yelyang 2015). Times of "lights off" made bathing to deal with irregular symptoms challenging. Storage of water for later bathing use when there was no piped supply became important.

These experiences illustrate the need for researchers and WASH practitioners to pay attention to bathing. Geographical and social discourses provide insight into cultural perspectives of bathing in the Global North, which view bathing as a means to be accepted in society (Pickerill 2015). Bathing is seen as providing three benefits. Sanitation and social order denotes membership of civil society. Hydrotherapy and gentility marks high social status. Comfort, cleanliness and commodification address concerns of image and appearance (Shove 2003). The clear needs of perimenopausal women suggest the practical aspects of bathing, especially in low-income locations, also require attention.

Laundry

Laundry is another aspect of WASH that is rarely mentioned. Different symptoms of the perimenopause lead to an increase in laundry. Women change their clothing more frequently as clothes and bed sheets became stained and soiled due to sweating and menstrual blood flow:

Experiencing nights and day sweats means you will change your clothes often so that increased my laundry...I change twice or three times in a day.

(Mansa, aged 57, Kotei)

Sometimes when I go to sleep and wake up in the morning I realize my bedspread is soiled with blood. That makes me wash all the time.

(Abla, aged 50, Kotei)

Physical infrastructure services for washing clothes in La and Kotei are generally inadequate. Laundering is part of women's daily domestic duties, but symptoms of the perimenopause impede the ability of women to wash their clothes. Women hand-washed clothes, using a bowl of water and soap, which became more challenging due to joint pains during the perimenopause and beyond. Women needed to adapt by finding comfortable positions to sit in to make this less painful.

Heavy menstrual bleeding impacted laundry processes. Group discussions with women in participatory mapping sessions identified that women set aside separate underwear for when they were menstruating heavily. Blood stains on underwear and menstrual cloths were a source of embarrassment, even when laundering, and women were wary, hiding blood stains from the sight of others:

What we do is that we wash the cloth so the foam comes on top of the water to cover up the blood.

(Felicia, aged 69, La)

Menstrual blood is socially taboo in male-dominated Ghanaian society. Women are considered unclean during their menses and may not take part in social activities and, notably, must stay away from men (Bhakta et al. 2016). Laundry conventionally takes place in compounds where other people, including men, are in the vicinity. Laundry during the perimenopause was more heavily soiled due to the heavier menstrual flow than during conventional menstruation. Whilst the laundering was done where it could be seen, women used the physical bubbles of the soapy water to conceal the stains within their metaphorical bubbles in which they experience the perimenopause. Water acts as a physical barrier to hide and cleanse away increased menstrual blood.

Hidden reproductive geographies: Learning from the WASH needs of perimenopausal women

This chapter has demonstrated perimenopausal experiences can be indirectly understood through observing women's practical interactions with the built environment. Women cannot easily describe their experiences of the perimenopause to others, but we can read their use of WASH infrastructure to convey their tacit messages about the perimenopause. Yet, WASH practices during the perimenopause are conducted in spaces such as latrines that are not visible to others. The hidden geographies of the perimenopause lie buried in private spaces, although these are at times within public settings, particularly in countries such as Ghana. The practices also have a temporal aspect, with some activities taking place at night, when public WASH services may be inaccessible. The repeated 24-hour cycle of night sweats and other symptoms is a significant factor in the narrative of women's experiences, in contrast to the longer, more linear pattern of narratives about birth.

As women's bodies become unpredictable and ever changing, so too do their MHM practices. With the onset of very heavy menstruation, women deal with multiple hygiene management issues privately, often at night. Washing bodies and clothes stained with blood in turn stain the water. Inadequate sanitation infrastructure can expose the taboo, private symptoms of the perimenopause in a public space. Private concerns are exposed by ill-designed spaces such as public sanitation facilities. Laundry, bathing, water supply and sanitation shape the physical and social experiences of women through the perimenopause.

Examining the experiences of perimenopausal women expands existing debates on gender-water geographies (Alhers and Zwarteveen 2009; Harris 2009; Sultana 2009;). By looking at the intimate hygiene needs of perimenopausal women, this chapter expands these debates by drawing upon women's own needs for, and individual relations with water. This chapter reinforces Laws' (1993) argument that patriarchal structures determine women's spatial mobility, as perimenopausal women's movements are influenced by infrastructure built by male providers.

Summary

The study of the perimenopause has been absent from reproductive geographies. The experiences and needs of perimenopausal women are hidden, exemplified through an absence of literature, and are individually experienced by each woman who does not discuss her needs with others. The post-2015 development agenda, through its Sustainable Development Goals, highlights the importance of paying attention to women's sexual and reproductive health, however, the perimenopause remains absent from existing discussions on women's sexual health in development, geography and beyond.

Perimenopausal women's needs are hard to study directly but can be understood through a built environment perspective, specifically through the entry point of their experiences as users of WASH. Water and sanitation provision in countries such as Ghana is not ideal, but exploring perimenopausal women's specific WASH needs reveals insights into the geographies of the perimenopause as a reproductive life stage. In turn, the breadth of exploring the geographies of reproduction is expanded from the start of the reproductive cycle, reflected in geographies of birth and breastfeeding, to the end, the perimenopause.

Note

1 All names have been changed to protect the identities of the women.

Acknowledgements

Our thanks go to all of the women who have kindly shared their stories for this research, to Loughborough University for funding the study, and the wider research team.

References

Adinkrah, Mensah. 2017. *Witchcraft, Witches and Violence in Ghana*. New York: Berghahn.

Alhers, Rhodante and Zwarteveen, Margaret. 2009. "The Water Question in Feminism: Water Control and Gender Inequities in a Neo-liberal Era." *Gender, Place and Culture* 16(4): 409–426

Amaro, Hortensia. 1995. "Love, Sex, and Power. Considering Women's Realities in HIV Prevention". *American Psychologist* 50(6): 437–447.

Avis, Nancy E., Alicia Colvin, Joyce T. Bromberger, Rachel Hess, Karen A. Matthews, Marcia Ory, and Miriam Schocken. 2009. "Change in Health-related Quality of Life over the Menopausal Transition in a Multiethnic Cohort of Middle-aged Women: Study of Women's Health Across the Nation". *Menopause* 16(5): 860–869.

Awuah, Esi, Samuel F. Gyasi, Helen Apina, and Kweku Sekyiamah. 2014. "Assessment of Rainwater Harvesting as a Supplement to Domestic Water Supply: Case Study in Kotei-Ghana". *International Research Journal on Public and Environmental Health* 1(6): 126–131.

Bairy Laxminarayana, Shalini Adiga, Parvathi Bhat and Rajeshwari Bhat. 2009. "Prevalence of Menopausal Symptoms and Quality of Life after Menopause in Women from South India". *Australia and New Zealand Journal of Obstetrics and Gynaecology* 49(1): 106–109.

Bayor, Isaac and Albert Yelyang. 2015. "The Ghana 'Dumsor' Energy Setbacks and Sensitivities: From Confrontation to Collaboration". *West Africa Network for Peacebuilding Warn Policy Brief, Ghana*. March 15http://www.wanep.org/wanep/index.php?option=com_content&view=article&id=735:policy-brief-on-ghana-the-ghana-dumsor-energy-setbacks-and-sensitivities-from-confrontation-to-collaboration&catid=25:news-releases&Itemid=8.

Belcher, D. W. 1974. "Male and Female Family Planning KA.P. data". Paper read at Danfa project review meeting Ghana Medical School.

Bhakta, Amita, Julie Fisher and Brian Reed. 2014. "WASH for Perimenopause in Low-income Countries: Changing Women, Concealed Knowledge?" in *Proceedings of the 37th WEDC Conference, September 15–19*. Loughborough: WEDC, Loughborough University.

Bhakta, Amita, Gloria Annan, Yvonne Esseku, Harold Esseku, Julie Fisher, Benjamin Lartey and Brian Reed. 2016. "Finding Hidden Knowledge in WASH: Effective Methods for Exploring the Needs of Perimenopausal Women in Ghana", in

Proceedings of the 39th WEDC Conference, July 15–19. Loughborough: WEDC, Loughborough University.

Bhakta, Amita, Brian Reed and Julie Fisher. 2017. "Cleansing in Hidden Spaces: The Bathing Needs of Perimenopausal Women" in *Proceedings of the 40th WEDC Conference, July 24–28.* Loughborough: WEDC, Loughborough University.

Boot, M., and Sandy Cairncross. 1993. *Actions Speak: The Study of Hygiene Behaviour in Water and Sanitation Projects.* The Hague: IRC International Water and Sanitation Centre and London: The London School of Hygiene and Tropical Medicine.

Boyer, Kate. 2012. "Affect, Corporeality and the Limits of Belonging: Breastfeeding in Public in the Contemporary UK". *Health & Place* 18(3): 552–560.

Boyer, Kate. 2009 "Of Care and Commodities: Breast Milk and the New Politics of Mobile Biosubstances" *Progress in Human Geography* 34(1): 5–20.

Braun, Bruce. 2002. *The Intemperate Rainforest.* Minneapolis: Minnesota University Press.

Campioni, M. 1997. "Revolting Women: Women in Revolt" in *Reinterpreting Menopause: Cultural and Philosophical Issues* edited by Paul Komesaroff, Phillipa Rothfield, and Jeanne Daly, 77–99. London: Routledge.

Castelo-Branco, Camil, Santiago Palacios, Desiree Mostajo, Cristina Tobar, and Sueli von Helde. 2005. "Menopausal Transition in Movima Women, a Bolivian Native-American." *Maturitas* 51(4): 380–385.

Chambers, Robert. 1997. *Whose Reality Counts? Putting the First Last*London: Intermediate Technology Publications.

Chompootweep, S., M. Tankeyoon, K. Yamarat, P. Poomsuwan., and N. Dusitsin. 1993. "The Menopausal Age and Climacteric Complaints in Thai Women in Bangkok." *Maturitas,* 17(1): 63–71.

Crofts, Tracey and Julie Fisher. 2012. "Menstrual Hygiene In Ugandan Schools: An Investigation Of Low-Cost Sanitary Pads" *Journal of Water, Sanitation and Hygiene for Development,* 2(1) : 50–58.

Da Silva, A.R. and d'Andretta Tanaka, A.C. 2013. "Factors Associated with Menopausal Symptom Severity in Middle-aged Brazilian Women from the Brazilian Western Amazon" *Maturitas,* 76(1): 64–69.

DeLyser, Dydia, and Wendy Shaw. 2013. "For Menopause Geographies" *Area* 45(4): 504–506.

Duckitt, Kirsten. 2010. "Managing Perimenopausal Menorrhagia" *Maturitas* 66(3): 251–256.

Engineers without Borders. 2017. "Engineers without Borders", Last modified 2017, http://www.ewbchallenge.org/content/aims-objectives

Field, M. 1960. *Search for Security: An Ethnopsychiatric Study of Rural Ghana.* London: Faber and Faber.

Fine, Michelle. 2008. "An Epilogue… of Sorts" in *Revolutionising Education: Youth Participatory Action Research in Motion* edited by Julio Cammarota and Michelle Fine, 213–234. London: Routledge.

Fisher, Julie. 2008. "Women in Water Supply, Sanitation and Hygiene Programmes". *Proceedings of the ICE: Municipal Engineer,* 161(4): 223–229.

Fisher, Julie, Sue Cavill and Brian Reed. 2017. "Mainstreaming Gender in the WASH Sector: Dilution or Distillation?" *Gender and Development* 25(2): 185–204.

Gadalla, F. 1986. "Social, Cultural and Biological Factors Associated with Menopause". *Egyptian Journal of Community Medicine* 2(1): 49–62.

Gagnon, Anita, Lisa Merry, Jacqueline Bocking, Ellen Rosenberg and Jacqueline Oxman-Martinez. 2010. "South Asian migrant Women and HIV/STIs: Knowledge, Attitudes and Practices and the Role of Sexual Power". *Health and Place* 16 (1): 10–15.

Ghana Statistical Service. 2014. *Population and Housing Census: District Analytical Report – La Dade-Kotopon Municipal Assembly*, Ghana Statistical Service, Ghana.

Gough, Katherine and Paul Yankson. 2011. "A Neglected Aspect of the Housing Market: The Caretakers of Peri-Urban Accra, Ghana" *Urban Studies* 48(4): 793–810.

Greer, Germaine. 1991. *The Change: Women, Aging and the Menopause.* London: Hamish Hamilton.

Harris, Leila. 2009. "Gender and Emergent Water Governance: Comparative Overview of Neoliberalized Natures and Gender Dimensions of Privatization, Devolution and Marketization". *Gender, Place and Culture* 16(4): 387–408.

House, Sarah, Suzanne Ferron, Marni Sommer, and Sue Cavill. 2014. *Violence, Gender and WASH Toolkit: A Practitioner's Guide.* Available online at: http://violence-wash.lboro.ac.uk/ [Accessed: 5/5/2016].

Jewitt, Sarah and Ryley, Harriet. 2014. "It's A Girl Thing: Menstruation, School Attendance, Spatial Mobility and Wider Gender Inequalities in Kenya". *Geoforum* 56(1): 137–147.

Johnson, B. C. 1982. "Traditional Practices Affecting the Health of Women in Nigeria" in *Traditional Practices Affecting the Health of Women and Children* edited by T. Baasher, R. H. Bannersman, H. Rushwan, et al., 23-25. Alexandria, Egypt: World Health Organisation.

Jones, John Paul, Heidi J. Nast, and Susan M. Roberts. 1997. *Thresholds in Feminist Geography: Difference, Methodology, Representation.* Lanham, Maryland: Rowman and Littlefield Publishers.

Kidney, Maria, Linden Edgell and Friends of Londiani. 2013. "Girls for Girls Programme, Kenya" in *Proceedings of the 36th WEDC conference,* Nakuru, Kenya July 1–5.

Kindon, Sara, Rachel Pain, and Michael Kesby. 2007. *Participatory Action Research Approaches and Methods: Connecting People, Participation and Place.* London: Routledge.

Lartey, Benjamin. 2016. Personal communication. 20 January 2016.

Laws, Glenda. 1993. "Women's Life Course, Spatial Mobility and State Politics", in *Thresholds in Feminist Geography: Difference, Methodology, Representation* edited by John Paul Jones, Heidi J. Nast, Susan M. Roberts, 47–64.

Laws, Glenda. 1995. "Theorising ageism: lessons from postmodernism and feminism". *The Gerontologist* 35(1): 112–118.

Leathes, Bill. 2012. *Topic Brief: Delegated Management of Water and Sanitation Services in Urban Areas: Experiences from Kumasi, Ghana.* Available online at: http://www.wsup.com/resource/delegated-management-of-water-and-sanitation-services-in-urban-areas-experiences-from-kumasi-ghana/ [Accessed: 20/05/2016]

Linde, Charlotte. 2001. "Narrative and Social Tacit Knowledge". *Journal of Knowledge Management* 5(2): 160–170.

Loue, Sana and Martha Sajatovic. 2008. *Encyclopaedia of Ageing and Health.* New York: Springer Science and Business Media.

Luft, Joseph. 1961. "The Johari Window: A Graphic Model of Awareness in Interpersonal Relations" *NTL Human Relations Training News* 5(1): 6–7.

MacKian, Sara. 2008. "What the Papers Say: Reading Therapeutic Landscapes of Women's Health and Empowerment in Uganda", *Health and Place* 14(1): 106–115.

Mackie, Fiona. 1997. "The Left Hand of the Goddess: The Silencing of Menopause as a Bodily Experience of Transition", in *Reinterpreting Menopause: Cultural and Philosophical Issues* edited by Paul Komesaroff, Philippa Rothfield and Jeanne Daly, 17–31London: Routledge.

Moore, Ben. 1981. "Climacteric Symptoms in an African Community" *Maturitas*, 3 (1): 25–29.

Nonaka, Ikujiro and Hiro Takeuchi. 1995. *The Knowledge-Creating Company: How Japanese Companies Create the Dynamics of Innovation.* London: Oxford University Press.

Nukunya, G.K. 1969. *Kinship and Marriage among the Anlo Ewe.* London: The Athlone Press.

O'Neill, Maggie. 1996. "Researching Prostitution and Violence: Feminist Praxis" in *Women, Violence and Male Power* edited by Marianne Hester, Liz Kelly, and Jill Radford, 130–147Buckingham: Open University Press.

Pickerill, Jenny. 2015. "Cold Comfort? Reconceiving the Practices of Bathing in British Self-Build Eco-Homes", *Annals of Association of American Geographers*, 105(5): 1061–1077.

Prior, Jerilynn C. 2005. "Clearing Confusion About Perimenopause". *British Columbia Medical Journal* 47(10): 538–542.

Reed, Brian J. and Jane Bevan. 2014. *Managing Hygiene Promotion in WASH Programmes* available online at: http://wedc.lboro.ac.uk/resources/booklets/G013-Hygiene-promotion-online.pdf [Accessed 28/9/2017].

Robson, Elspeth. 2006. "The 'Kitchen' as Women's Space in Rural Hausaland, Northern Nigeria", *Gender Place and Culture* 13(6): 669–676.

Rogers, Wendy. 1997. "Sources of Abjection in Western Responses to Menopause" in *Reinterpreting Menopause: Cultural and Philosophical Issues* edited by Paul Komesaroff, Philippa Rothfield and Jeanne Daly, 225–238. London: Routledge.

Sampselle, Carolyn M., Siobán D. Harlow, Joan Skurnick, Linda Brubaker and Irina Bondarenko. 2002. "Urinary Incontinence Predictors and Life Impact in Ethnically Diverse Perimenopausal Women" *Obstetrics and Gynecology* 100(6): 1230–1238.

Shenton, Andrew K. 2007. "Viewing Information Needs Through a Johari Window". *Reference Services Review* 35(3): 437–496.

Shove, Elizabeth. 2003. *Comfort, Cleanliness and Convenience: The Social Organisation of Normality.* Oxford: Berg.

Sommer, Marni. 2011. "Integrating Menstrual Hygiene Management (MHM) into the School Water, Sanitation and Hygiene Agenda" in *Proceedings of the 35th WEDC Conference*, July 6–8 2011. Loughborough: WEDC, Loughborough University.

Stanley, Liz and Sue Wise. 1983. *Breaking Out.* London: Routledge.

Stearns, Vered, Lynda Ullmer, Juan F. Lopez, Yolanda Smith, Claudine Isaacs, and Daniel F. Hayes. 2002. "Hot Flushes". *Lancet* 360(9348): 1851–1861.

Sultana, Farhana. 2009. "Community and Participation in Water Resources Management: Gendering and Naturing Development Debates from Bangladesh". *Transactions of the Institute of British Geographers* 34(3): 346–363.

Torpy, Janet M., Cassio Lynm, and Richard Glass. 2003. "Perimenopause: Beginning of Menopause", *Journal of the American Medical Association*, 289(7): 940.

Tucker, Andrew. 2016. "Sexual Health: Section Introduction" in *The Routledge Research Companion to the Geographies of Sex and Sexualities*, edited by Gavin Brown and Kath Browne, 257–262. Oxon: Routledge.

UN (United Nations). 2015. *World Population Prospects: The 2015 Revision*. New York: United Nations.

UNICEF and WHO (United Nations Children's Fund and World Health Organisation). 2015. *Progress on Sanitation and Drinking Water 2015 Update and MDG Assessment*. Geneva: UNICEF and WHO.

Utian, Wulf. 1999. "The International Menopause Society: Menopause-related Terminology Definitions". *Climacteric* 2(4): 284–286.

Wambua, L.T. 1997. "African Perceptions and Myths about Menopause". *East African Medical Journal* 74(10): 645–646.

Wasti, S., S. C. Robinson, Y. Akhtar, S. Khan, and N Badaruddin. 1993. "Characteristics of Menopause in Three Socioeconomic Urban Groups in Karachi, Pakistan". *Maturitas* 16(1): 61–69.

WHO/UNICEF (World Health Organisation and United Nations Children's Fund). 2017. *Progress on Drinking Water, Sanitation and Hygiene Update and SDG Baselines* Geneva: UNICEF and WHO.

WSSCC and FANSA. 2015. *Leave No One Behind: Voices of Women, Adolescent Girls, Elderly and Disabled People, and Sanitation Workers*. Geneva: WSSCC and Freshwater Action Network South Asia (FANSA).

Part II
Places

5 "Here we are!"

Exploring academic spaces of pregnant graduate students

Katie Merkle

She was very active at that stage and I felt her a lot. So, while I was presenting I felt her move and I know it's visible. I don't know if they could see it, but I imagine an arm, a leg sticking out. It was so on two levels. Like, I remember thinking, can they [dissertation committee] see this happening? And then for me, just knowing it's happening, and trying to stay focused while this child is moving inside of me. I was like, here we are!

– Helen, Ph.D. in Women's Studies

Introduction

In the excerpt above, Helen, who has since completed her Ph.D. in Women's Studies, describes defending her dissertation proposal while eight months pregnant. Helen's description elegantly captures what it feels like to be pregnant during such a defining moment in graduate school. Part of what makes her defence story so captivating is how she describes experiencing a dual bodily existence. When Helen says, "Here we are!" she is signalling that she sees her body as more than just herself—that she sees herself as two bodies. During her presentation, Helen's body held space as a graduate student and as a pregnant person, whose internal movements were not entirely her own. Additionally, Helen's description illustrates how, for her, the proposal defence and pregnancy were not mutually exclusive. These two bodily experiences cannot be uncoupled; they are forever connected through time and place. Helen's pregnancy influenced her graduate school experience and being a graduate student affected how Helen experienced her pregnancy.

Pregnancy is a unique time of embodied change and growth. The body swells, fluids are expelled and then a baby is born. In conjunction with bodily changes, pregnancy can also usher in new embodied identities. Pregnancy, breastfeeding and motherhood can all fall under the umbrella of maternal bodies and identities (Longhurst 2008; Gatrell 2013). By listing pregnancy, breastfeeding and motherhood in this order, maternal identities can be thought of as a journey. This journey is also a gendered journey as maternal identities are gendered identities and maternal bodies are gendered bodies.

The social construction of gender cannot be detached from the social construction of maternities. As a result, it should come as no surprise then that many feminist scholars have analysed maternal bodies and identity formation.

Within feminist scholarship on maternities, motherhood receives the most attention from feminist scholars (e.g., Rich 1976; O'Reilly 2010; Kawash 2011) and "Motherhood Studies" is now a well-respected sub-field within the discipline of Women's and Gender Studies (Kawash 2011). Feminist geographers expand feminist scholarship by arguing that gender, sexuality and sex—as symbolic social constructions, performances and social relationships—connect to place and representations of place (McDowell 1999; Johnston and Longhurst 2010). Feminist geographers therefore examine motherhood and maternities through a spatial lens. Research ranges from breastfeeding to childbirth preferences (e.g., Pain et al. 2001; Boyer 2010, 2011, 2012; Emple and Hazen 2014). At the centre of this work is an examination of how pregnancy, childbirth and motherhood are embedded in a set of social conditions that can be manipulated based on the gender politics that are continually culturally reproduced (Longhurst 2008).

The purpose of this chapter is to advance feminist geographic scholarship on maternities by exploring how pregnancy is situated and materialised in place. Feminist geographers examine the social and political inequalities that exist within the gender binary and analyse how concepts of gender relate to space, place and social power (McDowell 1999; Johnson 2000; Longhurst 2008). In this chapter, I examine the lived embodied experiences of pregnancy of 20 graduate students at colleges and universities throughout the United States. Based on these embodied experiences, I argue that pregnant graduate students are positioned as bodies *out of place* and *on display* because their pregnant bodies and identities transgress what is considered normal bodily behaviour on college and university campuses.

Feminist geography and maternal bodies

Geographers have long been concerned with how people experience, access, position and situate themselves within place. Places are not static backdrops; rather, people live their lives through and within place, as places are constructed by the "physical, historical, social, and cultural" (Casey 2001, 683). Places are meaningful and powerful, as their meanings and power reflect and contribute to societal formations. Such constructions of meaning and power influence both what sort of behaviours are acceptable where and who "belongs in one place, but not in another" (Cresswell 1996, 3).

Ideas about what are considered appropriate behaviours are embedded within place. Place influences who has power, or lacks power, and where. Cresswell (1996, 11) explains that the metaphorical expression "out of place" is used "to clearly demarcate whether social or geographical place is denoted—place always means both". Bodies feel and are positioned as "out of place" when they transgress what is considered acceptable normal behaviour within a particular place. The notion of transgression implies the crossing of a normative boundary connected to hegemonic ideologies, again reinforcing the spatial metaphor. Cresswell (1996, 9) explains that transgression, "serves to foreground the mapping of ideology onto space and place, and thus, the margins can tell us something about normality". In this context, different bodies experience place in different ways, as factors such as gender, race, ability and other corporeal markers influence what types of bodies belong where (Fluri and Trauger 2011). As McDowell (1999, 4) notes:

> Places are made through power relations, which construct the rules which define boundaries. These boundaries are both social and spatial—they define who belongs to a place and who may be excluded, as well as the location or site of the experience.

Gender, along with gendered power and gendered subjectivities, is culturally produced and maintained through everyday interactions within everyday places. Therefore, representations of place contribute to how power and subjectivity develop and maintain gendered power relations.

Feminist geographers illuminate these power relations that people experience in their everyday lives by focusing on gendered relationships in everyday places. Looking at how everyday places are gendered helps to highlight and deconstruct the politics that lie behind gender relations (Domosh and Seager 2001).

Within the context of feminist geography, where research on the body and embodiment continues to grow, the body is defined as a material site where personal identities are built and actualised (Bell et al. 2001; Johnston and Longhurst 2014). Bodies can be both privileged and Othered depending on the spaces they inhabit. Furthermore, feminist geographers position the body as a surface of social, cultural and political inscription and, therefore, a site where power and subjectivity are constructed. Looking at and examining how the body or a group of bodies function in particular places serves as a way to explore the relationships between gender, place, power and subjectivity (Johnston and Longhurst 2010). Politics, specifically gender politics, produce the spaces that make up everyday geographies, and this construction influences the diversity and complexity of maternal bodies in relation to identities and experiences (Longhurst 2008).

Feminist geographic scholarship on maternal bodies highlights the way that maternal bodies are strongly gendered, resulting in women feeling out of place in public spaces during pregnancy and lactation. Feminist geographer, Robyn Longhurst, for example examines how spatiality impacts how first-time pregnant women felt. In Longhurst's research, pregnant participants express feelings of emotional discomfort in public places. Longhurst theorises such feelings stem from the messy materiality of pregnant bodies. Under this framework, maternal bodies disturb social order by changing in shape and size as well as with their production and emission of fluids. Bodily fluids are not merely the stuff of our biology, they also influence social and gender relations. Leakages are often met with feelings of revulsion, which reflects the "theoretical concept of abjection, in which bodily fluids are seen to threaten social norms" (Gatrell 2013, 624). Maternal bodily fluids are not just indicators of biological differences; they also illuminate how bodily boundaries are culturally produced by hegemonic patriarchal structures (Longhurst 2008).

Methodology

All research is partial, and this study is grounded in the perspectives of the graduate student participants. I conducted this research using critical feminist qualitative research methods. Feminist methodology seeks to be inclusive and ethical and does not seek to universalise experiences or identities. In fact, it celebrates diversity while exploring and trying to understand better the multiple, and many times contradictory, truths that surround human lives (Moss 2002). Furthermore, this approach to research critiques and challenges social structures that harm and abuse people. Qualitative research methods are commonly used within human geography because they allow for "multiple conceptual approaches and methods of inquiry" (Winchester 2010, 3). I asked questions surrounding the intersection of social and institutional power structures and embodied experiences. The specific methods for this research were designed to gain a better understanding of what life is like while pregnant in graduate school. I relied exclusively on semi-structured in-depth interviews because semi-structured interviews are content-focused and allow for greater flexibility and fluidity between the researcher and participant. This research is pushing back on privileging the masculine position by engaging in research with methods that ethically illuminate experiences while also challenging the gendered politics that surround knowledge production.

Participants

I conducted 20 interviews with graduate students between January and March 2016. Demographically, this group of interviewees was very homogeneous. All participants identified as cisgender women, 18 participants identified as white, one as black and one as brown, one interviewee was an international student, and the rest were U.S. citizens. The participants were, however, diverse regarding their level of education, academic discipline, age when they experienced pregnancy and where they attended graduate school in the United States. All the women had experienced pregnancy while pursuing a full-time graduate degree between 2013 and 2016; four students were still pregnant at the time of our interview (Figure 5.1). Participants attended colleges and universities from the southern, northwestern, midwestern, and northeastern regions of the United States. The majority of graduate students attended state universities. Interviews occurred through video chat via Skype or Facetime, except for two, which took place in person at a public place of the participant's choosing.

I selected particular university and community organisations to recruit participants. I then reached out by email, public announcements and print advertisements—all of which explained the research objectives, personal compensation and how to contact me. The interviews lasted from 60 to 120 minutes, and questions focused on lived experiences and personal reflections of what pregnancy was like while attending graduate school. All interview participants received a $15.00 Visa gift card for their time. In this chapter, all participants are identified by a pseudonym.

Since this research is also interested in institutional policy and procedures that affect pregnant bodies at universities, I also conducted three semi-structured interviews with university professionals: a programme coordinator within a Women's Centre, a graduate student academic advisor and a faculty member who served as the advisor to board members on the graduate student senate. While these interviews are not the focus of this chapter, they nonetheless provided contextual information which I used to inform the analysis.

Lack of maternal bodies and maternal representations

All participants in this study noted that visibly pregnant people are generally not seen on their college or university campus. Kira, a Ph.D. candidate in Women's and Gender Studies, told me, "visibly pregnant women are just absent from the public space of the university. I definitely felt like I stood out like a sore thumb." Kira's experience of feeling out of place reflects the more general trend within the context of academic spaces, where women's bodies are not perceived as the norm. As Britton (2000) and Acker (1990) argue, the generic ideal of the worker is based on universal notions rooted in a masculine ideal. Within the contexts of particular workspaces, feminist geographers such as McDowell (1999) have argued that workers and workspaces are not

---

Graduate Students Interviewees

14 doctoral candidates / students

- 2 participants were pregnant during our interview
- 12 participants were pregnant between 2013 and 2016
- 4 participants had a natural sciences discipline as their home discipline
- 10 participants had a social sciences or humanities discipline as their home discipline

6 Master's students

- 2 participants were pregnant during our interview
- 4 participants were pregnant between 2013 and 2016
- 2 participants had a natural sciences discipline as their home discipline
- 4 participants had a social sciences or humanities discipline as their home discipline

Figure 5.1 Academic discipline characteristics of graduate student participants

gender neutral. Rather, she argues that "organizations reflect masculine values and men's power, permeating all aspects of the workplace in often taken-for-granted ways" (McDowell 1999, 136). The experiences of maternal bodies within academia are no different; as Castaneda and Isgro (2013, 3) state, "the ideal intellectual worker in the academy is male gendered".

Feminist scholars are beginning to look at the hardships and difficulties that come from combining working as an academic with being a mother (e.g., Raddon 2002; Evans and Grant 2008; Castaneda and Isgro 2013; Mason et al. 2013). Such difficulties illuminate the unequal gender representations regarding professionals working in the academy (Evans and Grant 2008). Far more men than women have positions of power and prestige within academia (Mason et al. 2013), even though women receive slightly more than half of the doctoral degrees granted in the United States. More women than men work in lower ranking positions, such as being a part-time faculty member. In regard to parenthood status, women who work in lower ranking positions are as likely as their male counterparts to be parents. However, women who hold top positions within the academy are far less likely to be married or have children than their male counterparts (Mason et al. 2013). These examples indicate the way academic work continues to be gendered masculine.

Moreover, the pressures placed on academics to "be productive" trickle down and influence the lives of graduate students. Graduate student participants reported feeling as if they always needed to be in production mode. The pressure to produce as much as possible impacted how participants viewed pregnancy. Many participants believed their academic departments saw pregnancy and motherhood as potential interferences for possible production. Dauphine, who had her first child while completing her Ph.D. in Biology, explained how disruptions to production can be viewed negatively within academia:

> [My son] came two weeks early, which is in the realm of pretty normal, but I was just trying to get as far as I could with my research. So, literally the day before he was born I did a surgery on one of my research animals and I taught a class. I was really tired and I was planning to do a surgery the next day and I was going to meet [my advisor], and I just didn't feel right when I woke up, and of course it turns out I'm in labour, so I worked literally till the night before he was born. I went in and had him and within the next two weeks I turned in three grants, and when he was two-and-a-half weeks old I got an email from my advisor that said, "You've missed [colloquium]", which is like this weekly seminar. "I wasn't aware this was optional".

Dauphine's story illustrates both the level of pressure she was under to be productive and her perception that anything that interferes with a student's productivity—in this case pregnancy and motherhood—would be viewed negatively. Even though we don't know the motivations of her advisor, it is significant that Dauphine attributed the negative treatment she received from her advisor to the fact that she had just had a baby.

According to many of my research participants, it is essential for a graduate student to have a supportive and attentive advisor for them to be successful in their programme (Hawkins et al. 2014). However, this relationship between advisor and advisee is riddled with complex power dynamics that typically place all the power in the hands of the advisor (Hawkins et al. 2014); these inequalities are then socially reproduced at both institutional and inter-personal levels (Freeman 2000). If an advising relationship sours or is not productive, it can drastically affect the academic progression of a graduate student. Graduate students lack institutional power and interpersonal power with their advisors (Hawkins et al. 2014). Many of my graduate student interviewees believed that pregnancy was a possible reason for causing an advising relationship to turn from productive to counterproductive, as they believed that pregnancy was seen—by their advisors and their academic departments—as something that would limit academic production. Such a situation happened to Tiffany, a Ph.D. candidate in Women's and Gender Studies:

> Because I was pregnant, I felt a lot of pressure to try and finish every-thing up so that I would be more present for my child. During that phase I was like writing my proposal for my dissertation and, I mean, I wouldn't get any feedback from [my advisor] for like months. It would take [my advisor] three months to respond back. Yeah, it was crazy. And in fact, that semester of my dissertation, which was my pregnancy period from my second to third trimester, I think I gave her a draft in like October and I never got it back until after my daughter died, and my daughter died in January, two days after she was born. It was just a frustrating semester because here I am, trying to move forward. It just was really hurtful to me. It took that very drastic thing to happen in my life for her to be like, "Well, here is your draft." All semester long I've been telling you I want my draft back so I could be present. It was a very low blow to me and I literally didn't do anything on it, I didn't respond for a couple months later 'cause I was just trying to get over my daugh-ter's death.

Tiffany's loss is heartbreaking. During our conversation, we talked a lot about her pregnancy, her daughter and her daughter's death. Tiffany's experiences and feelings show how pregnancy can have the potential to affect the working relationship between a graduate student and their advisor. Tiffany's and Dauphine's observations that their advisors treated them poorly because of their pregnancies signal a transgression of social and spatial boundaries which code what is normal and expected within academic spaces.

These fears manifested themselves when it came time for participants to tell their advisors they were pregnant. Many participants noted that disclosing their pregnancy to their academic advisor(s) was a nerve-racking experience. Participants were scared to reveal their pregnancies because they did not want to be viewed as less serious and they did not want to "disappoint" their advisors. Allison, a Master's student in Sociology, who was pregnant with her first child during our interview, describes sending an email to her advisor about her pregnancy. She comments:

> When I was writing the email to my advisor I didn't realise it was going to be scary till I hit send. Then I was like checking my phone every few seconds just to make sure she wasn't going to kick me out of the program because I was pregnant. Which is absurd, she was happy for me. But it was a lot more stressful than I thought it would be...After sending it, I thought, "What if she is not supportive?" So I was just kind of doubting myself. I thought about having an extra responsibility and still having to do all the things that it takes to be successful in grad school. Like maybe she would think I can't do all of them cause it's like a whole extra thing. Like she is really successful and doesn't have kids. I think that part of my stress was coming from that. Just not knowing how much, not like sympathy or empathy, but how much understanding she would have was difficult to gauge.

Allison's fear that she would be "kicked out" of her programme and university illustrates how pregnancy was seen as something which did not belong or was "out of place" within the university. The rarity of pregnancies on campus reinforced respondent's feelings of being "out of place" on campus. All of my graduate student participants stated they were the only pregnant student in their department or had been the only pregnant student for some time. Ashley, a Ph.D. candidate in Evolutionary Biology who was pregnant with her first child, noted of her department:

> Only one other [female] graduate student in recent history has had a child and there are a couple post-docs with kids. One is a male, but really very, very few. Out of how many people we have, we have like 30 just in our group within the department and there are not that many people with kids.

This lack of pregnant bodies on campus caused a broad range of experiences for my graduate student participants. For instance, Anna, a Ph.D. candidate in Religion who had one of her two children while in graduate school, explained that because she was the only pregnant person in her programme she became reference material for other graduate students in her programme. Anna stated:

> I feel like I've kind of become a totem pole or something for whenever a graduate student is even thinking about it [becoming pregnant or having

a child]. They will find me and ask me what was it like, even if they don't know me all that well. It's like they are scoping out first-hand experience with me as to whether they are going to try it out. It's like they are curious. They are thinking about it.

Anna further explained to me how she thoroughly enjoyed discussing her pregnancies and experiences of motherhood with her fellow graduate students. However, the enjoyment Anna experienced in those conversations does not diminish the fact that she was an outlier in her department. Moreover, Anna illustrates how the pregnant body is seen as open to the public as she became a point of reference on pregnancy for other graduate students, even for those she did not know well. In this respect, Anna's body was not only "out of place" but also "on display"—a topic I turn to in the next section.

The pregnant body: On display and out of place

Many graduate student participants reported feeling "out of place" and "uncomfortable" in public places where they were not known on their university campuses. In particular, many respondents reported feeling self-conscious while walking across campus. Walking through public spaces invoked feelings of separation, vulnerability, difference, isolation and an overall sense of discomfort. Experiencing these feelings caused many participants to avoid public spaces in general. The reason for this discomfort, according to the participants, related to feelings of wanting to be viewed by the public as a "legitimate" and "respectable" body. For example, Kira explains why she felt uncomfortable walking around her campus:

So, I'll have to give you background on this. I have another child who was born while I was in undergrad and I placed him for adoption, for an open adoption. My college pregnancy was an extremely stigmatising and traumatising experience. You know my whole family flipped out. I had to leave undergrad. It was just considered by everyone to be this huge failure on my part. And I ultimately ended up leaving my son because of a lot of things but mainly because of the social stigma around me being a single, young, unwed woman from a conservative Catholic family. So that was my first experience with pregnancy on campus, long before I went to grad school. And so as a grad student, and an instructor and advisor, and I happen to be someone who looks younger than my age, I would be walking from the parking garage to my office and I would just, like, have these moments. I wondered if people looking at me thought I was an undergrad and pregnant. It was only because I felt that if people thought I was an undergrad and pregnant, this is bad, but they are looking at me like I'm a slut or a whore. And there was no way on the walk for me to be like, "I'm an adult!"

Kira's experience of pregnancy in graduate school was unique as it connected her back to her undergraduate experience. She was not the only participant who was uncomfortable at the thought of being mistaken for an undergraduate however. Ashley (Evolutionary Biology) echoes Kira's concerns:

> I've been kind of glad that it is winter actually. Largely because I can wear a bigger coat and people don't necessarily know [that I am pregnant], but other times it is super obvious when I'm walking around from class to class. Like, I know I put this on myself, but I feel awkward and kind of like judged or something because I know they don't know my situation or anything, and I hope I look a little bit older than the undergrads. I don't know. It makes me feel weird 'cause people stare at me. They will be walking and then I walk by and they follow me with their eyes.

This notion of being a body "out of place" directly connects to the politics that surround place and gender within colleges and universities (McDowell 1999). Pregnancy here transgresses a social boundary of respectability and legitimacy.

The classroom serves as another specific place within colleges and university which evoked feelings of being on display. Madeline, an M.A. student who was very visibly pregnant at the time of our conversation, was enrolled in a dual graduate and undergraduate class during her pregnancy. When I asked her about her experiences as a student in a classroom setting, she stated:

> The first day of class, [other students in the class] were actively staring at my belly. It was just, like, mortifying. So, it's kind of been an interesting experience. Especially in, like, these 50-minute classes. You don't have a lot of time to talk to people. You don't really get to know anyone. I have noticed, like, a definite difference in how they view my pregnant body versus how people in my programme do. I mean some [people in my programme] went straight from undergrad to grad but most people have been out working for a while, so they have, like, friends who are pregnant or married, or sisters or brothers. These undergrad students are, like, totally weirded out by a pregnant person in their class.

Madeline's statement further demonstrates how pregnant bodies are viewed and internalised differently depending upon their spatiality. These feelings of vulnerability correlate to literature that argues that because pregnant bodies challenge bodily and social boundaries, they lack social power in public places (Longhurst 2008). Power, or the lack thereof, within this context was associated with not feeling like a "legitimate" and responsible body within the place of the classroom. Being a pregnant undergraduate student would not be viewed as responsible, thus the participants did not want to be associated with such an identity.

Graduate students in natural sciences disciplines sometimes reported feeling like they were viewed as biological specimens during their pregnancy. They had the sense that their bodies were being looked at and studied from a biological/scientific perspective by fellow graduate students, from faculty and staff within their departments and by students in their classrooms. This sentiment comes out very clearly in the following comment from Ashley (Evolutionary Biology) as she discusses teaching an anatomy class:

> [Fellow graduate students] have been really intrigued because I think they are scientists and they have no filter as far as asking me, like, "what's going on right now?" and "what is the weirdest thing that's happening to me?" and then the medical students are a whole other can of worms. Some of them caught on, probably late October, and so they were asking, like, a lot of personal questions. Which is OK with me. I teach anatomy so I have to talk about all the gross stuff all the time. So it didn't really faze me, but I just thought it was funny how open they felt asking me certain questions. You know as we went through the uterus and looking at the pelvis and that sort of thing. [The medical students] are, like, "Oh what's going on with your uterus?"

Dauphine explained a similar experience of feeling on display amongst her colleagues:

> I had this ligament in my hip that was just really sore, so my waddle was very pronounced some days because it was just so painful to walk, and I worked on this hallway with this gait cycle biomechanist [a person who researches how humans walk] who was, like, analysing how I walked and she was, like, you really need to try not to do that and of course my response was, like, you've never been pregnant!

Dauphine's and Ashley's pregnant bodies were on display because they were positioned as out of place. Their pregnant bodies received public attention because they transgressed what a normative body looked like within those spaces.

Conclusion

This chapter explored how pregnant graduate students experience the places that make up American colleges and universities, including the classroom and open public spaces, through a critical feminist geographic framework. As the participants expressed, their feelings about their body changed depending on where they were located spatially. To this end, this research illuminated how social boundaries and embodied identities correlate to the places and spaces that surround everyday academic geographies. Furthermore, this study contributes to broader scholarship on maternities by exploring the intersections of pregnancy and institutions from the disadvantaged position of a graduate student.

The research presented here demonstrates how pregnant graduate students felt "out of place", not only in their university (in general) but also in its actual spaces. These graduate students were fearful they would be forced to leave their university when they disclosed their pregnancy. Participants felt vulnerable, out of place and on display when in places they were not well known, like public spaces and classrooms. They felt safe and secure when they were in places on campus where they were well known. Graduate students also described feeling that their pregnant bodies were on display within the classroom and in their academic departments. Graduate students who were studying in natural science departments, in particular, reported feeling as if their bodies were on display for study as biological specimens. The experiences and fears expressed by participants illustrate how pregnancy is deemed as something which does not belong on college and university campuses.

I concluded every interview by asking participants what advice they would give to a graduate student who just found out they were pregnant. All participants expressed, in some way, that one should be proud of their pregnancy, and I would like to end with Helen's advice. She states:

> Be unapologetic about it. I feel especially in academia you almost have to apologise for being pregnant or it's, like, expected to hide, but if you're just unapologetic about being pregnant like I feel like for me, it's, like, created situations where they don't feel like they can confront me on it. I am confident in this pregnancy and it's not something that I am ashamed of or going to hide. So when you are unapologetic about it, people have no choice but to accept it. Like, this is how it's going to be.

Helen's quotation is important for two reasons. First, by implying there may be a need or feeling that one should apologise for their pregnancy, it illustrates how pregnancy in graduate school transgresses constructed expectations of graduate students; it breaks a social boundary. Second, Helen's statement is empowering. Not only is she critiquing the current system but she is trying to encourage the development of a new one—a landscape where pregnant graduates do not feel out of place, where they feel comfortable and confident taking up space within their college and universities.

Acknowledgements

I am forever indebted to the 20 graduate students for sharing their experiences with me. Thank you for sharing your stories. I'd also like to thank Marcia England, Helen Hazen, Maria Fannin and Risa Whitson for their comments and support as I developed this chapter.

References

Acker, Joan. 1990. "Hierarchies, Jobs, Bodies: A Theory of Gendered Organizations." *Gender & Society*, 4(2), 139–158.

Bell, David, and Jon Binnie, and Ruth Holliday, Robyn Longhurst, and Robin Peace. 2001. *Pleasure Zones: Bodies Cities Space*. Syracuse, New York: Syracuse University Press.

Boyer, Kate. 2010. "Of Care and Commodities: Breast Milk and The New Politics of Mobile Bio-Substances." *Progress in Human Geography 34(1)*: 5–20.

Boyer, Kate. 2011. "The Way to Break the Taboo Is to Do the Taboo Thing' Breastfeeding In Public and Citizen activism in the UK." *Health and Place 17(2)*: 430–437.

Boyer, Kate. 2012. "Affect, Corporeality, and The Limits of Belonging: Breastfeeding in Public in the Contemporary UK." *Health and Place 18(3)*: 552–560.

Britton, Dana M. 2000. "The epistemology of the gendered organization." *Gender & Society*, 14(3): 418–434.

Butler, Judith. 1990. *Gender Trouble: Feminism and the Subversion of Identity*. New York, NY: Routledge.

Butler, Judith. 1993. *Bodies That Matter: On the Discursive Limits of Sex*. New York, NY: Routledge.

Caputo, John, and Mark Yount. 2006. *Foucault and the Critique of Institutions*. University Park, PA: Penn State University Press.

Casey, Edward. 2001. "Between Geography and Philosophy: What Does It Mean to Be in the Place-World?" *Annals of the Association of American Geographers 91(4)*: 683–693.

Castaneda, Mari, and Kristen Isgro. 2013. *Mothers in Academia*. New York, NY: Columbia University Press.

Creswell, John W. 1998. *Qualitative Inquiry and Research Design: Choosing Among Five Traditions*. Thousand Oaks, CA: Sage Publications.

Cresswell, Tim. 1996. *In Place/Out of Place: Geography, Ideology, and Transgression*. Minneapolis, MN: University of Minnesota Press.

De Beauvoir, Simone. 2011. *The Second Sex*. New York, NY: First Vintage Books.

Domosh, Mona, and Joni Seager. 2001. *Putting Women in Place*. New York, NY: The Guilford Press.

Elwood, Sarah A., and Deborah G. Martin. 2000. "'Placing' Interviews: Location and Scales of Power in Qualitative Research." *Professional Geographer 52(4)*: 649–657.

Emple, Hannah, and Helen Hazen. 2014. "Navigating Risk in Minnesota's Birth Landscape: Care Providers' Perspectives." *ACME: An International E-Journal for Critical Geographies 13(2)*: 352–371.

Evans, Elrena, and Caroline Grant. 2008. *Mama Ph.D.* Piscataway, NJ: Rutgers University Press.

Fox, Rebecca, and Kristen Heffernan, and Paula Nicolson. 2009."'I Don't Think It Was Such a Big Deal Back Then': Changing Experiences of Pregnancy Across Two Generations of Women in South-East England." *Gender, Place, and Culture 16(5)*: 553–568.

Fluri, Jennifer, and Amy Trauger. 2011. "The Corporeal Maker Project (CMP): Teaching About Bodily Difference Identity and Place Through Experience." *Journal of Geography in Higher Education 35(4)*: 551–563.

Foucault, Michel. 1995. *Discipline and Punish*. Translated by Alan Sheridan. New York, NY: Random House.

Frank, Arthur. 1990. "Bringing Bodies Back In: A Decade Review." *Theory, Culture and Society* 7: 131–162.

Fraser, Gertrude. 1995. "Modern Bodies, Modern Minds: Midwifery and Reproduction Change In An African Community." In *Conceiving the New World Order*, edited by Fay Ginsburg and Rayna Rapp, 59–74. London, England: University of California Press.

Freeman, Amy. 2000. "The Spaces of Graduate Student Labor: The Times for A New Union." *Antipode 32(3)*: 245–259.

Gatrell, Caroline J. 2013. Maternal Body Work: How Women Managers and Professionals Negotiate Pregnancy and New Motherhood at Work. *Human Relations*, 66 (5): 621–644.

Grosz, Elizabeth. 1995. *Space, Time, and Perversion: Essays on the Politics of Bodies.* New York, NY: Routledge.

Guba, Egon, and Yvonna S. Lincoln. 1998. "Competing Paradigms in Qualitative Research." In *The Landscape of Qualitative Research: Theories and Issues,* edited by Norman K. Denzin and Yvonna S. Lincoln, 195–220. Thousand Oaks, CA: Sage Publications.

Hartman, Yvonne, and Sandy Darab. 2012. "A Call for Slow Scholarship: A Case Study on the Intensification of Academic Life and Its Implications for Policy." *Review of Education, Pedagogy, and Cultural Studies 34(1)*: 49–60.

Hawkins, Roberta, Maya Manzi, and Diana Ojeda. Lives in the Making: Power, Academia and the Everyday." *ACME: An International E-Journal for Critical Geographies 13*, no. *2(2014)*: 328–351.

Johnson, Candace. 2014. *Maternal Transition: A North-South Politics of Pregnancy and Childbirth.* New York, NY: Routledge.

Johnson, Louise. 2000. *Place Bound: Australian Feminist Geographies.* South Melbourne, Victoria, Australia: Oxford University Press.

Johnston, Lynda, and Robyn Longhurst. 2010. *Space, Place, and Sex: Geographies of Sexualities.* Plymouth, UK: Rowman & Littlefield.

Kawash, Samina. 2011. "New Directions in Motherhood Studies." *Signs: Journal of Women in Cultural and Society 36(4)*: 970–1003.

LeCompte, Maragaret, and Jean J. Schensul. 1999. *Designing and Conducting Ethnographic Research.* Walnut Creek, CA: Altamira Press.

Lefebvre, Henri. 1991. *The Production of Space.* Translated by Donald Nicholson-Smith. Oxford: UK.

Longhurst, Robyn. 1995. "Viewpoint The body and Geography." *Gender, Place and Culture 2(1)*: 97–106.

Longhurst, Robyn. 2001. *Bodies: Exploring Fluid Boundaries.* New York, NY: Routledge.

Longhurst, Robyn. 2008. *Maternities: Gender, Bodies and Space.* New York, NY: Routledge.

Mason, Mary Ann, and Nicholas Wolfinger, and Marc Goulden. 2013. *Do Babies Matter? Gender and Family in the Ivory Tower.* New Brunswick, NJ: Rutgers University Press.

McDowell, Linda. 1999. *Gender, Space and Place.* Cambridge, UK: Polity Press.

Meyerhoff, Eli, and Elizabeth Johnson, and Bruce Braun. 2011."Time and the University". *ACME: An International E-Journal for Critical Geographies 10(4)*: 483–507.

Moss, Pamela. 2002. *Feminist Geography in Practice.* Oxford, UK: Blackwell Publishers.

Mullings, Beverely, Linda Peake, and Kate Parizeau. 2016. "Cultivating an Ethic of Wellness in Geography." *The Canadian Geographer 60(2)*: 161–167.

Nelson, Lise. 1999. "Bodies (and Spaces) do Matter: The Limits of Performativity." *Gender, Place & Culture 6(4)*: 331–353.

O'Neil, John, and Patricia Leyland Kaufert. 1995. "Irniktakpunga!: Sex Determination and the Inuit Struggle for Birthing Rights in Northern Canada". In *Conceiving the New World Order,* edited by Fay Ginsburg and Rayna Rapp, 59–74. London, England: University of California Press.

O'Reilly, Andrea. 2010. "Outlaw(ing) Motherhood: A Theory and Politic of Maternal Empowerment for the Twenty-first Century." *Hecate 36*: 17–29.

Pain R., Bailey, and G Mowl. 2001. "Infant Feeding in North East England: Contested Spaces of Reproduction." *Area 33(3)*: 261–272.

Raddon, Arwen. 2002. "Mothering in the Academy: Positioned and Positioning Within Discourses of the 'Successful Academic' and 'Good Mother.'" *Studies in Higher Education 27(4)*: 387–403.

Rich, Adrienne. 1976. *Of Woman Born: Motherhood as Experience and Institution.* New York, NY: Norton.

Rossman, Gretchen, and Sharon F. Rallis. 2003. *Learning in the Field: An Introduction to Qualitative Research.* Thousand Oaks, CA: Sage Publishers.

Smith, Neil. 2000."Who Rules the Sausage Factory?" *Antipode 32(3)*: 330–339.

Smith, Sarah. 2016. "Intimacy and Angst in the Field." *Gender, Place & Culture 23(1)*: 134–146.

Whitson, Risa. 2007. "Hidden Struggles: Spaces of Power and Resistance in Informal Work in Urban Argentina." *Environment and Planning 39(12)*: 2916–2934.

Winchester, Hilary. 2010. "Qualitative Research and its Place in Human Geography." In *Qualitative Research Methods in Human Geography,* edited by Iain Hay, 3–24. New York, NY: Oxford University Press.

Young, Iris Marion. 2005. *On Female Body Experience: "Throwing Like a Girl" And Other Essays.* New York, NY: Oxford University Press.

6 "It is a jail which does not let us be…"

Negotiating spaces of commercial surrogacy by reproductive labourers in India

Dalia Bhattacharjee

Introduction

"A fertile ground for exploitation of women" bears the headline for *The Hindu*, one of the leading Indian newspapers (Dhar 2012). Articles showing similar headlines have been very popular in Indian mainstream media, referring to several aspects of commercial surrogacy that need attention. These concerns include: the amount of money this thriving industry makes, the exploitation that women working as surrogate mothers face, the misinformation fed to the women regarding the medical process, violations of workers' rights and the inadequate legal framework guiding commercial surrogacy. However, such discussions draw attention away from the experiences of the women working as surrogate mothers themselves. The voices of reproductive labourers are overshadowed by the dominant narrative of victimhood. Such an absence has been normalised in Indian society as bearing children outside the heterosexual institution of marriage is not considered morally acceptable and, hence, is silenced.

My research questions the invisibility of the women who remain at the centre, and yet at the periphery, of commercial surrogacy arrangements in India. While the "wombs-for-rent" industry has been flourishing, the authentic voices of the women performing this role are hidden behind the dominant narrative of exploitation of women's bodies and the "un-naturalisation" of motherhood. This chapter reports the narratives of women working as surrogate mothers and argues for the inclusion of commercial surrogacy within the larger ambit of "work" so that the women are not merely recognised as "surrogate" or "replaced" mothers. Consistent with this approach, I call the women *reproductive labourers* rather than surrogate mothers, proposing that this broader term downplays the one-sided discourse of victimhood and acknowledges the ways in which the women re-position themselves as active labourers in the exchange market.

It is important to note that in downplaying notions of victimhood, my approach by no means attempts to glorify commercial surrogacy. Given the industry's improper functioning in India, associated with government negligence towards the rights and interests of the reproductive labourers, the industry remains troubling. In particular, the government has been negligent in ensuring adequate payment for reproductive labourers (currently only four lakh rupees, around U.S. $6,100, for one term), in formulating a regulatory framework governing surrogacy services and in providing health insurance to women working as reproductive labourers. Nonetheless, I argue that many women "choose" to become surrogate mothers within the limited avenues available to them, and that it is appropriate to consider the ways in which women use this arena to provide opportunities to earn money for the betterment of their families and in turn to celebrate their self-worth. In order to better understand women's decisions around reproductive labour, I focus on analysing the links between production and reproduction and in challenging the dichotomy between the two.

Commercial surrogacy in India

Commercial surrogacy has been a topic of debate in Europe and America since the 1980s. However, it is a comparatively new phenomenon in the Global South. With the growing demand for such services, it has come to be a multibillion-dollar industry, with India—up until recently—one of the most popular locations. However, in 2016 the Indian government banned commercial surrogacy in the country, allowing only altruistic surrogacy for married Indian heterosexual couples. Such a ban poses a setback to the industry in India. This complicates the situation further with every possibility of the industry being pushed underground.

Debates around commercial surrogacy have been emerging from various perspectives, ranging from liberal, to legal, to ethical; with many scholars highlighting the exploitative nature of the practice and debating ethical aspects of such services (Andrews 1987; Arneson 1992; Raymond 1993; Ragone 1994; Bailey 2011). There has also been some recent ethnographic scholarship on commercial surrogacy that emphasises the lived experiences of women working as reproductive labourers in India (Pande 2014; Rudrappa 2015; Vora 2015). Although this turn is much needed and acknowledged, there is still a dearth of such ethnographies. In this chapter, I extend this recent path towards documenting the experiences of reproductive labourers in order to elucidate the relationship between reproduction and labour, and to highlight the ways in which reproductive labourers navigate and re-work disruptions in the meanings of motherhood and work.

I aim to understand the ways in which reproductive labourers experience motherhood at the intersection of the public and private, or work and non-work. This study maintains that commercial surrogacy subverts such gendered dichotomies, but at the same time also mirrors them. For instance, while reproduction has crossed out of the boundary of the home, the surrogate housing facility functioning in India claims to provide a private and homelike environment for the women and yet simultaneously denies the women living there any privacy.[1] This chapter argues that it is not enough to reveal the unsteadiness of the public/private binary, but instead urges us to consider what such a blurring does to the reproductive labourers in terms of empowering or disempowering them.

Within these discussions, reproductive labourers are also at risk of being labelled "immoral women", as their reproductive activities transgress the boundaries of the private sphere. A second major theme of this chapter is the ways in which reproductive labourers navigate this cultural context by equating surrogacy with virtuous work that empowers them to help people. The narratives presented in this chapter reveal a rejection of the societal portrayal of reproductive labourers as "immoral" or "dirty" by stressing their special role as reproducers. My respondents also often distinguish surrogacy from sex work by describing their work as pious. It is in this manner that reproductive labourers negate the stigma attached to them by tacitly and creatively questioning the images that society assigns them.

The research focuses on fieldwork undertaken at a surrogate house in the city of Anand, in the state of Gujarat, where reproductive labourers stay from the time their bodies are being prepared for pregnancy till they give birth. Reproductive labourers are hired by the medical authorities through the signing of a legal contract between them and the commissioning couple. The contract withholds the right of the reproductive labourer to the baby soon after it is born and fixes the amount to be paid to her. She is then kept in the hostel where her body is made biologically ready to conceive. The women are made to take medications and meals on time, as prescribed by doctors and dieticians, and may not eat their own choice of food. They cannot move out of the surrogate house; nor are they allowed to be sexually involved till the baby is born. The centre has a house matron who oversees the activities of the women. Additionally, their emotional well-being is controlled, with the goal that they should not feel any attachment to the babies they carry. The women are repeatedly reminded that they are merely "carriers" of the babies, which belong to other people.

Any attachment to the baby that a reproductive labourer develops may lead to an adulteration of the role she is supposed to play as gestational mother and reproductive labourer, making her a corrupt worker who does not perform her duties to her full potential. Instead, her role is simply to deliver a baby to a childless couple and earn money for the betterment of her own children. It is these factors which make the infertility clinic produce the surrogate body in such a manner that it works to take care of the baby like a dutiful mother till it is born, and yet does not develop any emotional bond so that, like a responsible worker, the reproductive labourer can give away the "product" as soon as it is ready.

As long as the reproductive labourer adheres to these rules, she is considered to be a good worker. However, the moment she transgresses these boundaries, she is open to the risk of being labelled a "bad worker" or even a "bad mother". In this way, commercial surrogacy acts as a challenge to the age-old dichotomies between production and reproduction, public and private. It pushes reproduction and childbirth outside of the so-called privacy of the patriarchal heterosexual family and into the ever patriarchal and repressive capitalist market, where women's reproductive capacities are valued and monetised (Boris and Parreñas 2010; Cooper and Waldby 2014; Pande 2014).

Methodology

Anand is popularly identified as India's "surrogacy capital" (Bhalla and Thapliyal 2013), as well as a global surrogacy hub. The city is also known more broadly for medical tourism, with a number of hospitals, health centres, medical agencies, pharmacies, private nursing homes and clinics set up in the city in the last couple of decades. In addition, high-class hotels have been established that can accommodate medical tourists from across the world.

I refer to the site for this study using the pseudonym: Nishaan Infertility Clinic. Dr. Sudha Vinayak (also a pseudonym), the clinical director of the Nishaan Infertility Clinic, is known as one of the best IVF specialists in India and has an international reputation as a fertility specialist. The surrogate housing facility where I undertook my research is housed in the same building as the hospital facilities associated with the clinic.

The chapter is informed by qualitative research methods. The data come from ethnographic fieldwork over a span of six months and three additional visits during the period 2015–2017. All interviews were recorded in Hindi with the consent of the respondents and translated and transcribed in English. The respondents include: 40 reproductive labourers, three intended parents, two doctors (including the medical director), one nurse and the hostel matron. Of the 40 labourers, 15 were repeat reproductive labourers and 25 were women who had newly joined the surrogate house. The women who agree to work as reproductive labourers are all aged between 25 and 35. This clinic allows women to work as reproductive labourers a maximum of three times, with a minimum of two years between pregnancies, depending on the woman's reproductive health and age. All the reproductive labourers interviewed came from villages and cities near Anand. Most of the commissioning couples whom I interviewed were from outside India (mainly from the developed world) or were non-resident Indians (NRIs) living abroad.

The Nishaan Infertility Clinic underwent rapid development between my first visit in 2014 and my latest visit in early 2017. Originally located in the busiest marketplace of the city as a small clinic, the facility is now being transformed into a multi-storey five-star hospital on the outskirts of the city. It has also been renamed, using the title "research institute", and now provides services beyond reproductive treatments, including dentistry, cosmetic surgery and physiotherapy. Previously, the surrogate house was separate from the clinic (about 5 kilometres away), but the surrogate housing facility has now been moved to the basement of the hospital building.

The new building has a centralised air conditioning system and large rooms for the reproductive labourers, with each room occupied by four women. Each room has a big-screen LCD television and an attached washroom with a hot water tank. There is also a laundry facility and a cafeteria with catering services for the reproductive labourers. The food is prepared in accordance with the instructions of a dietician, who comes to monitor the food and the women as they consume their meals. No outside food is allowed in the housing. The facility has a big common space with tables and chairs, which doubles up as the dining area. One corner has an altar with pictures of Hindu and Christian gods and goddesses. The lives of the reproductive labourers are run according to the timetable fixed by Dr. Vinayak and the activities of the reproductive labourers are monitored by an elderly woman, who lives on site as matron.

Challenging the public/private dichotomy

Several scholars have shown how industrialisation and domesticity in the nineteenth and twentieth centuries in Western societies contributed to a divided city along gender lines (Yeoh and Huang 1998). In such a city, the suburb signifies the private sphere, or the domain of women and "non-work", while the central city is defined as the public sphere, dominated by men and "productive" work (McDowell 1983; Miller 1983; Pateman 1989). Women's reproductive labour has always been confined to the family or the private sphere, and hence not considered labour at all but a naturalised aspect of her being. In this chapter, I argue that the consideration of commercial surrogacy as non-work is based on the dichotomies of public/private, production/reproduction, culture/nature and work/non-work. Such distinctions count as the central theme around which the feminist movement has taken shape (Pateman 1989). Feminist scholars argue against such divides so that the subordination of women in the labour market can be challenged (McDowell 1991; Hanson 1992; Rose 1993; Hanson and Pratt 1995).

Anthropologist Michelle Rosaldo (1974) states that the distinction between private and public draws a clear-cut line that excludes much work done by women, including childcare, housework and so on. She argues that such boundaries can lead to the subordination of women as they are confined to the domestic sphere and excluded from the public. Reproduction and motherhood remain affairs guided by the patriarchal principles of the private spaces of the home. Feminist scholars like Adrienne Rich (1995), Andrea O'Reilly (2010), Carole Pateman (1988) and Sara Ruddick (1995), among others, see motherhood as a key site of oppression of women in a patriarchal structure. Thus, motherhood has often been taken as the pre-eminent characteristic that disqualifies women from most activities of public life.

The idealisation of motherhood as sacrosanct and holy in India and elsewhere considers the mother as the nurturer and naturalises the reproductive labour performed by her, hence normalising and routinising women's efforts in the process of childbirth. In a patriarchal society, the mothering body is produced in such a manner that it distances itself from monetary value. Such discourses and practices place motherhood away from the public sphere, so that there is no room for its association with work or politics (Lane 2014). However, with the commodification of motherhood and reproduction, the reproducing body gains importance in the exchange market. Women working as surrogate mothers are now viewed as reproductive machines and are described as "incubators" (Hollinger 1984, 901), "rented wombs" (Corea 1985, 222) or "surrogate uteruses" (Burfoot 1988, 110). This shift has added a non-human and profitable aspect to some parts of women's bodies (ova, womb) in the capitalist economy. Such a change has also facilitated a shift in the experiences of motherhood from the private space of the home to the public space of the infertility clinic and the cover pages of newspaper discussions.

Surrogacy as hard labour

Based on my ethnographic journey in a fertility clinic in India, I approach commercial surrogacy as a form of work, which is physical as well as emotional in nature. Such a perspective aims to unveil and acknowledge the hard labour performed by reproductive labourers. As Maria Mies (1986) argues, the sexual division of labour must be challenged so that women's child-bearing and child-rearing capacities are perceived as conscious social acts and as work. In this direction, I use the work of the Anglo-American philosopher Donna Dickenson (2007), who argues for the need to acknowledge women's labour in pregnancy and childbirth, which has been overshadowed by the perspective of such labour being seen merely as renting out wombs. Drawing on Marx's theory of alienation, Dickenson (2007, 54) asserts,

> The alienated worker's labour is always in fact the symbol of his oppression, not of his freedom, although under capitalism he is not a slave. In the capitalist system, writes Marx, labour is none the less external and forced, even though the labourer is not physically compelled to work, as the slave is. But at least Marx credits the worker with a property in his own labour, which is more than women have in relation to the new reproductive technologies.

In her work, Dickenson calls for the need to recognise women's property rights to their ova so that their efforts in the ova market can be made visible. She asserts that women's reproductive labour has been reduced to a natural function, which remains the most significant reason behind the neglect of their hard work in reproducing. Along similar lines, I argue that there is an emergent need to destabilise the naturalness of women's reproductive work in order to make women's efforts visible in commercial surrogacy. My respondents frequently emphasised the economic role that they were playing in their families, in a form of resistance to this naturalisation of their reproductive function. Pooja, a 27-year-old reproductive labourer, notes:

> I think about surrogacy as a kind of service. It's like any other job that a person does to earn money. Like a man sells his labour while driving, or carrying weight, or even while teaching. Giving birth in surrogacy is also a service. It's not an easy job and not anyone can do it.

Subscribing to the view that surrogacy is a form of work, women like Pooja strengthen their importance in the labour market by challenging the division of labour based on one's sex.

While highlighting their significant role in the labour market, the reproductive labourers also clearly lay out their economic significance in the household owing to payments from surrogacy work. Rani, a 31-year old woman, was hired by a couple from the U.S. Rani's husband works in agriculture and earns only a meagre amount of money, insufficient for the sustenance of the family. Rani has dreams of getting her eight-year old daughter a good education, which she feels can be fulfilled through surrogacy. She remarks:

> I feel so proud of myself. My daughter is going to get the best education ever. All the work that I am doing here is for her. The money is for her future. I am not getting my daughter married early. She'll also do a Ph.D. like you. My husband could never give her what she needed; but I can. I feel so strong and happy!

While women were frequently happy with the choice they had made to become reproductive labourers because of the economic boost it would provide to the family, the degree to which this reflects a genuine "choice" is questionable for women like Rani. Most of the reproductive labourers come from impoverished backgrounds in rural areas and have had little opportunity for education. They have husbands who work as agricultural labourers, construction workers, auto drivers or security guards. In such dire economic circumstances, the idea that women "choose" to take up surrogacy and leave the privacy of their own family to work in the public sphere as a reproductive labourer may be legitimately questioned. Nonetheless, women are clearly using their own agency to make the best future for their families in challenging circumstances.

Arguing that it is important that women's reproductive labour be viewed as a legitimate economic activity is also not to suggest that women get fair terms of trade. During my field visit in 2015, I witnessed a woman packing her bags and leaving the surrogate house with a heavy heart after she received a negative pregnancy report. When I asked her if she was given pay for the three months she had been there and undergone all the medications, she said, "Why will they pay me now? I am of no use to them. I have been told to leave this place". It is distressing that the labour performed by the women before a pregnancy starts is unpaid and ignored. As expressed by almost all the reproductive labourers I interviewed, that period is the most painful period for them, both physically and emotionally. They have to take the injections and medicines properly, often involving heavy doses, and their actions are overseen to the point of receiving instructions on what position to sit or lie in. One of my key respondents, Diksha, a 26-year-old reproductive labourer, shares her experiences:

> The process of [embryo] transfer was quite painful for me. Some of the women here find it very painful. I was like in a physical shock for quite some time after the process was over. Sometimes I cannot believe I am still alive. I am six months pregnant now. The initial three months were hell for me. Especially till the time I conceived. I developed a terribly sore back and it took me weeks to finally recover. It was the most painful experience ever.

In addition to the physical pain the women go through, the fear of going back home empty handed because of a failed pregnancy constantly haunts them. Sarita, a 27-year-old reproductive labourer from a nearby city, is working as a reproductive labourer in order to send her eight-year-old daughter to school. She says:

> I have promised my daughter that I'll come home only when I am able enough to send her to school. She is a very bright kid and wants to study like you. I am praying day and night that this implantation succeeds. I don't want to go back home and kill my daughter's dreams. This is the second embryo transfer. I hope I get pregnant this time. I have already spent more than a month here. They won't pay me anything if I fail.

Sarita's narrative of hope and courage is tempered by a fear of loss should her reproductive labour be considered to amount to nothing. I argue that such hard labour and the emotional toll on these women, even before conception, must be accounted for and included within the contract signed by the reproductive labourers and the commissioning couples.

In the case of commercial surrogacy in India, motherhood, reproduction and child-bearing—which essentially belong to the private space of the home—cross the boundary and enter into the public. This form of market economy blurs the distinction between reproduction and production, private and public. That this reality is so frequently obscured is, I argue, mainly because equating surrogacy with work would pose a challenge to the institution of motherhood, which is understood as sacrosanct and holy in India and supposed to function within the heterosexual private space of the home.

Surveilling the reproductive body in the surrogacy industry in India

Another way in which surrogacy upsets traditional notions of the public and the private is through the constant surveillance that the women must endure. Like any other capitalist structure, this labour-intensive market is always under scrutiny and constant monitoring. As identified by feminist scholars, the introduction and use of technologies and surveillance is characteristic of modern industry (Weedon 1987; Wright 2006). As witnessed in my research, the surrogate house acts as a closed monitored space, with the presence of the doctors, dietician and matron extending the "medical gaze" over the reproductive labourers in their care. The reproductive labourers inhabiting this space express a deep sense of dissatisfaction with the ways in which they are made to live under the vigilance of the authorities, as remarked by Pooja: "It is a jail which does not let us be. It's like our in-laws' place, where we are not free." This was expressed especially strongly by the repeat reproductive labourers, who contrasted the current arrangement with their experiences of staying in the older surrogate housing. According to them, the previous housing felt somewhat like home, while current facilities do not. As Pooja notes above, for many this situation equates to the experience of being in their conjugal homes where they have to always conform to their in-laws' wishes. Seema has served twice in the clinic. She compares her stay in both the housing facilities and notes:

> This new set up is worse. The old surrogate house was much more fun and interesting. We could sneak out of the house when we craved for outside food. Also, we could roam around a little and the doctor visits had fixed timings. Here, the doctors are present in the same building and come in anytime they want; outside food is out of the question here. You either eat or stay hungry. Adjusting here is no less than a challenge to us. It is no less than my in-laws' place.

In the previous set up, the women maintained some autonomy over their movements and activities. In the present system, by contrast, the women are only allowed to sit in the garden, which is inside the hospital premises. They are not allowed to visit their homes unless an emergency crops up in the family. The surveillance assures the disciplining of the bodies of the reproductive labourers and, in turn, guarantees the safety of the foetus that they carry. In such an environment, motherhood becomes a very stressful experience. The monitoring of the reproductive body in the surrogate housing makes sure that the body adheres to the rules and regulations laid down for it.

During my visits to the surrogate house, as per the instructions of Dr. Vinayak, I had to perform all interviews in the presence of a woman named Tina, who works as a counsellor in the hospital. She guides the new members in the group of reproductive labourers through the entire process and takes care of their needs. During one rare opportunity to talk to the labourers in

Tina's absence one day, I started a conversation about how I was missing my home in the eastern part of India. The conversation led to the unveiling of feelings of alienation among my respondents, associated with living in a space which is high-tech in nature but, according to them, is a space that cannot be considered home. In the course of the conversation, Deepa, a 29-year-old reproductive labourer, opens up:

> I feel lonely here. Though I have made some friends here, I feel sad and alone. I miss my home. They say this is my home, but it surely isn't. There is no privacy. We cannot even talk freely here as they might hear us. They might even lessen our pay. I don't know! I'm just counting days for this whole thing to get over so that I can get back to my children. I don't like this hi-fi place. I am better off at home.

The assurance of "home away from home" by the medical authorities is not upheld and the space fails to give a sense of privacy to the women who inhabit the surrogate housing. This sense of alienation experienced by the reproductive labourers informs us of the ways in which new technologies can monitor and harness the bodily capacities of the women who agree to carry babies for strangers.

Recognising commercial surrogacy as a form of labour leads to the disruption of the reproductive space of the home, which is private, and pushes reproductive activity into the public sphere, where it can be made productive. This disruption frequently leads to the critique that poor women are exploited at the hands of the techno-patriarchs of capitalism. While my interviewees undoubtedly had concerns over some of the conditions of their employment, as a group they denied this rhetoric of exploitation and wretchedness and instead emphasised their agency in challenging circumstances.

Stigma and status: The position of the reproductive labourer in Indian society

In moving the supposedly private act of reproduction into the public sphere, reproductive labourers are often viewed as immoral—either sleeping around or selling babies for money. Reproductive labourers are therefore frequently stigmatised in Indian society. Some of this is associated with misinformation. Most reproductive labourers come from rural areas and belong to the lowest economic and social strata of society. It is commonly believed in these communities that surrogacy involves sexual intercourse. Moreover, giving birth to children outside of marriage is also considered immoral.

My respondents, by contrast, drew a clear distinction between surrogacy and sex work, as surrogacy does not include any physical intimacy with the intended father because of the medical process behind the arrangement. Meenal, a 26-year-old woman hired by a couple from South Africa, states:

They say that I am not a good woman. I am selling the baby for money. But they don't know anything. I am doing a *punya ka kaam* [virtuous work]. We give children to the childless. We are gifted women. We are not like those women who sell their own children for money or sell their bodies. We are pious... and I am not genetically related to the children. I am not giving away my baby. It is theirs. I am just carrying it for them because she [the intended mother] has some problem. The child should go to the ones it belongs to and it will complete her [the intended mother's] family.

Meenal, like several other reproductive labourers, believes that what makes surrogacy different and morally correct is the completion of someone else's family, which counts as the ultimate and pious aim behind the reproductive labour that they perform. This narrative emphasises also how the absence of any genetic tie serves as a significant factor for the surrogate mother to defy the baby selling discourse.

For some women, their very womanhood was a gift that they perceived they could share with others. Basanti is a 30-year-old reproductive labourer who comes from Ahmedabad. Like several other reproductive labourers, she hid her employment from her in-laws. However, when they found out from another source that she was working as a reproductive labourer, they tried to force Basanti to come back to Ahmedabad. Nevertheless, Basanti stood by her decision and refused to go back home. She opines:

I am a woman. I am gifted with my reproductive capacity. Like an actor is gifted with his acting talent, a singer with his voice, I am gifted with this. If they can make money out of their capability, what's wrong when a woman wishes for the same? I don't accept such things. I am working for the betterment of my children and I'll do anything to improve their lives. This is no dirty work! But no one understands. I am not exploited here. I don't know how much money the doctors make, but I see this as the only way to make my children prosper and I am proud of what I do.

Most of the women I interviewed mentioned that they do not disclose to their friends and relatives that they work as surrogate mothers because of their fear of being ostracised by society. They fear that such stigmatisation will, in turn, have negative impacts on their children and other family members. Instead, they often tell people that they have been working as contract labourers in a distant city. Their absence from home also attaches the stigma of "absent mother" to them. This is especially significant in a cultural context, as "mother" is the core job that women are supposed to perform in India. Expressing her guilt from being away from her children, Seema, a 30-year-old reproductive labourer, states:

I feel really bad and guilty too. I am not being able to deliver my duties as a mother. My husband has to really struggle all the time in taking care of our children and doing his job. My in-laws have even said that I am not a good mother. But the work I am doing here, all the sacrifice and pain that I am going through, is for my children's bright future only. I wish I had also hidden this surrogacy thing from my in-laws like the other women. My family thinks I am a dirty woman.

The guilt of leaving their children back home with their husbands and not being able to take care of them, was experienced by most of the women I interviewed. Nevertheless, the reproductive labourers managed such feelings by emphasising the greater good that this sacrifice is going to bring in their lives. Furthermore, the women also reject the idea that they are neglecting their families by emphasising their worth as significant economic actors. For many reproductive labourers in India, commercial surrogacy has become a survival strategy for earning money to improve their own children's prospects, as noted by Seema above. The women take great pride in this work as they believe that it makes them even stronger than men, who are the traditional breadwinners in the family. Many women expressed how their status in their own families improved after returning to their homes with money to support the family.

Bringing surrogacy out of the purview of reproduction, which is presumed as holy, provides a lens to acknowledge the struggles of the reproductive labourers where they fight the identities assigned to them by Indian society as dirty and deviant. Many women emphasised the intrinsic value of the work they did, where their very womanhood provided a gift that they could share with others, while other women argued that the economic role they played was critical to the betterment of their household. I see these as resistive strategies used by reproductive labourers to downplay the dishonour and shame that society attaches to their activities.

Conclusion

Scholars have argued that public space normalises certain kinds of interactions and actions (masculine), which leads to the exclusion of other kinds of presence or identities that are considered problematic or deviant (not masculine or political) (Ruddick 1996; Valentine 1996). The exclusion of women from the public exchange market, and bracketing women's reproductive work as natural and private, remains at the core of their subordination. In this chapter, I have argued that the process of bearing and giving birth to the surrogate baby requires a great deal of intentionality and hard work. Reproductive labourers experience this labour as neither wholly productive, nor wholly reproductive. They create products of value for the exchange market which is public and hence productive. Yet, the process is reproductive in the sense that the product is extracted out of women's bodies in a clinic and not a factory. This fails to acknowledge the women as active workers in a capitalist market, but instead labels them as mere incubators in baby farms.

Reproductive labourers work 24/7 to give birth to the babies. This form of hard labour performed by reproductive labourers is also often stigmatised in society as dirty work or baby selling. Despite facing these forms of stigmatisation, women fight back with dignity, focusing on the larger aim of improving the financial stability of their families. This chapter has argued that the efforts made by reproductive labourers in the process of commercial surrogacy must be understood as work so that they are not re-essentialised within patriarchal conceptions of motherhood and women's work. The chapter also interprets commercial surrogacy as a form of intimate labour, which takes shape at the intersection of the public and private spheres.

The entire process of surrogacy in India results in alienation of the reproductive laborers from the "fruits of their labours" (i.e., the children born out of such contracts). Such an alienation has been an intrinsic part of the training process of the surrogate body in the IVF clinic. Using the narratives of reproductive labourers in this study, this chapter has demonstrated that this distancing is significant so that they are not perceived as unscrupulous workers or bad mothers. Women are also alienated from their families through staying in the surrogate house for the entire gestation period. Managing their feelings towards the unborn children they carry is nonetheless a significant challenge to reproductive labourers. They manage their emotions in such a manner that they are neither presented as desperate mothers who want to keep the babies, nor as baby sellers who give away the babies they birth in exchange for money. However, such sacrifice and hard labour performed by the women remains invisible in mainstream discourses around commercial surrogacy.

Surrogacy also results in the alienation of the reproductive labourers themselves, as this form of labour is gendered and, in India, highly stigmatised. Women use a variety of resistance strategies to reject such negative stereotypes through emphasising the important economic role they play and the way in which surrogacy allows them to use their own unique (female) gifts to provide a valuable service to another family.

Paying close attention to stratified reproduction, feminist theorists have discussed the ways in which developments in assisted reproductive technologies have been built upon power relations where some people are empowered to reproduce, while others are not. The Indian transnational commercial surrogacy industry has seen many of these disparities, where some women's reproductive capacity has been positioned above that of others. On the basis of these discussions and my own research, it can be said here that the global setting in which surrogacy takes place has been unequal, patriarchal and also displays instances of stratified reproduction. Based on the narratives of reproductive labourers in this study, commercial surrogacy is a remarkable instance of reproductive labour where production and reproduction intersect. Such an intersection shapes the entire procedure: the recruitment and disciplinary tactics used by the medical authorities, as well as the reproductive labourers' strategies of negotiating and resisting such a regime.

Note

1 I have visited infertility clinics in other cities in India, including Bangalore, Chandigarh and Delhi. I have witnessed similar surrogate housing facilities, or surrogacy centres, in all of them. The housing is not always in hospital buildings but is typically near the hospital and run by the same people who own the surrogacy centres. These people hire agents to bring women who agree to work as surrogate mothers to the centres. In some places, CCTV cameras have been installed in the rooms of the reproductive labourers to keep a close watch on their activities.

Acknowledgements

My gratitude goes to Dr. Anu Sabhlok, my Ph.D. supervisor, for her insightful comments and encouragement. My heartfelt thanks to the editors of the book Maria Fannin, Helen Hazen and Marcia England for their valuable comments and constructive feedback in shaping the chapter in its current form. A special thanks goes to my companion, Yogesh Mishra, for reading earlier drafts and sharing his views. This chapter is based on the paper presented at the AAG Annual Meeting, 2017, in Boston. For this, I am thankful to Indian Institute of Science Education and Research (IISER) Mohali (India) for providing financial support to attend the meeting and share my work.

References

Andrews, Lori B. 1987. "The Aftermath of Baby M: Proposed State Laws on Surrogate Motherhood." *Hastings Center Report* 17(5): 31–40.

Arneson, Richard J. 1992. "Commodification and Commercial Surrogacy." *Philosophy & Public Affairs* 21(2): 132–164.

Bailey, Alison. 2011. "Reconceiving Surrogacy: Toward a Reproductive Justice Account of Indian Surrogacy." *Hypatia* 26(4): 715–741.

Bhalla, Nita, and Mansi Thapliyal. 2013. India Seeks to Regulate its Booming 'Rent-a-womb' Industry. *Reuters*. September 30. Accessed December 21, 2013. http://www.reuters.com/article/us-india-surrogates/india-seeks-to-regulate-its-booming-rent-a-womb-industry-idUSBRE98T07F20130930

Boris, Eileen, and Rachel Parreñas. 2010. *Intimate Labors: Cultures, Technologies, and the Politics of Care.* California: Stanford University Press.

Burfoot, Annette. 1988. "A Review of the Third Annual Meeting of the European Society of Human Reproduction and Embryology," *Reproductive and Genetic Engineering* 1(1): 108–110.

Cooper, Melinda, and Catherine Waldby. 2014. *Clinical Labor: Tissue Donors and Research Subjects in the Global Bioeconomy.* Durham: Duke University Press.

Corea, Gena. 1985. *The Hidden Malpractice: How American Medicine Mistreats Women.* New York: Harper and Row.

Dhar, Aarti. 2012. A Fertile Ground for Exploitation of Women, Says Study. *The Hindu.* July 16. Accessed July 20, 2012. http://www.thehindu.com/todays-paper/tp-national/a-fertile-ground-for-exploitation-of-women-says-study/article3644005.ece

Dickenson, Donna. 2007. *Property in the Body.* UK: Cambridge University Press.

Dworkin, Andrea. 1978. *Right Wing Women*. New York: Perigee Trade.

Hanson, Susan. 1992. "Geography and Feminism: Worlds in Collision?" *Annals of the Association of American Geographers* 82(4): 569–586.

Hanson, Susan, and Geraldine Pratt. 1995. *Gender, Work, and Space*. Oxon: Routledge.

Hollinger, Joan Heifetz. 1984. "From Coitus to Commerce: Legal and Social Consequences of Noncoital Reproduction." *University Mich. JL Reform* 18: 865–932.

Lane, Rebecca. 2014. "Healthy Discretion? Breastfeeding and the Mutual Maintenance of Motherhood and Public Space." *Gender, Place and Culture* 21(2): 195–210.

McDowell, Linda. 1983. "Towards an Understanding of the Gender Division of Urban Space." *Environment and planning D: Society and Space* 1(1): 59–72.

McDowell, Linda. 1991. "Life without Father and Ford: The New Gender Order of Post- Fordism." *Transactions of the Institute of British Geographers* 16(4): 400–419.

Mies, Maria. 1986. *Patriarchy and Accumulation on a World Scale*. New York: Zed Books.

Miller, Raymond. 1983. "The Hoover in the Garden: Middle-class Women and Suburbanization, 1850–1920." *Environment and Planning D: Society and Space* 1: 73–87.

O'Reilly, Andrea. 2010. *Toni Morrison and Motherhood: A Politics of the Heart*. Albany: SUNY Press.

Pande, Amrita. 2014. *Wombs in Labor: Transnational Commercial Surrogacy in India*. USA: Columbia University Press.

Pateman, Carole. 1988. *Sexual Contract*. Redwood: Stanford University Press.

Pateman, Carole. 1989. *The Disorder of Women: Democracy, Feminism, and Political Theory*. Redwood: Stanford University Press.

Ragone, Helena. 1994. *Surrogate Motherhood: Conception in the Heart*. Boulder, CO: Westview.

Raymond, Janice G. 1993. *Women as Wombs: Reproductive Technologies and the Battle over Women's Freedom*. Australia: Spinifex Press.

Rich, Adrienne. 1995. *Of Woman Born: Motherhood as Experience and Institution*. New York: W.W. Norton & Company.

Rothman, Barbara Katz. 1988. "Cheap Labor: Sex, Class, Race—and 'surrogacy.'" *Society* 25(3): 21–23.

Rosaldo, Michelle Zimbalist. 1974. "Woman, Culture, and Society: A Theoretical Overview," In *Woman, Culture, and Society*, edited by Michelle Zimbalist Rosaldo, Louise Lamphere, 17–42. California: Stanford University Press.

Rose, Gillian. 1993. *Feminism & Geography: The Limits of Geographical Knowledge*. Minnesota: University of Minnesota Press.

Ruddick, Sara. 1995. *Maternal Thinking: Toward a Politics of Peace*. US: Beacon Press.

Ruddick, Susan. 1996. "Constructing Difference in Public Spaces: Race, Class, and Gender as Interlocking Systems." *Urban Geography* 17(2): 132–151.

Rudrappa, Sharmila. 2015. *Discounted Life: The Price of Global Surrogacy in India*. New York: N.Y.U. Press.

Valentine, Gill. 1996. "Lesbian Productions of Space," In *BodySpace: Destabilising Geographies of Gender and Sexuality*, edited by Nancy Duncan, 146–155. London: Routledge.

Vora, Kalindi. 2015. "Re-imagining Reproduction: Unsettling Metaphors in the History of Imperial Science and Commercial Surrogacy in India." *Somatechnics* 5(1): 88–103.

Weedon, Chris. 1987. *Feminist Practice and Poststructuralist Theory.* USA: Wiley-Blackwell.

Wright, Melissa W. 2006. *Disposable Women and other Myths of Global Capitalism.* USA: Taylor & Francis.

Yeoh, Brenda S.A. and Shirlena Huang. 1998. "Negotiating Public Space: Strategies and Styles of Migrant Female Domestic Workers in Singapore." *Urban studies* 35(3): 583–602.

7 The best of both worlds?

Mothers' narratives around birth centre experiences in the Twin Cities, Minnesota

Helen Hazen

Introduction

Many researchers have analysed birth geographically through consideration of the spatial expectations and implications of different places of birth (e.g., Jordan 1993; Sharpe 1999; Fannin 2004; Longhurst 2008; Cheyney 2011). These studies are united in their belief that place of birth is a critical point of investigation as it is intricately interwoven with power relations between birthing mothers and medical practitioners. These place-based power relations are constructed and reinforced by medical authority, as well as broader cultural understandings of birth. Given this power-laden framework around place of birth, it is important to interrogate and analyse changes in the birth landscape. This study considers the role that birth centres play in this highly contested and evolving birth landscape, using a case study from the Twin Cities, Minnesota, in the U.S. While birth centres are often presented as the "best of both worlds", combining the comforts of home with the perceived safety of the hospital, I argue that the true meaning of birth centres is more complex, with many birthing mothers viewing them as a compromise rather than an ideal. In this chapter, I situate mothers' experiences with birth centres in a broader discussion of the meaning of birth in the home and hospital, in order to explore the complexity of meanings associated with birth centres in the evolving birth landscape.

Although many hospitals now call their labour and delivery wards "birth centres", the strict sense of the term is normally reserved for a labour and delivery unit staffed by certified professional midwives that operates independently from the hospital setting. These so-called "freestanding birth centres" are able to operate outside the strict protocols of hospital medicine, although they adhere to their own professional standards and transfer patients to hospital in emergencies. A uniting feature of birth centres is a philosophical "commitment to normality of pregnancy and birth" (Laws et al. 2009, 290). This is usually operationalised via an emphasis on low-intervention births (for women with low-risk pregnancies). Although birth centres with this philosophy have existed in the U.S. for at least 50 years, there has been a significant resurgence in interest in out-of-hospital birth over the past 20 years, stimulating renewed interest in birth centres and the opening of many new facilities.

In the Western world, commentators often make a binary distinction between the hospital and the home as sites of birth, with dualisms such as masculine-feminine, medicalised-natural, obstetrician-midwife, and paternalism-autonomy identified with the hospital and home respectively (Sharpe 1999). In drawing these distinctions, emphasis is typically placed on hospitals as associated with obstetricians following a biomedical model of birth and home associated with midwives pursuing more holistic goals. Recent critical analyses of places of birth have critiqued such simplistic dualisms, arguing that consideration of home and hospital as places of birth requires a considerably more nuanced analysis than this binary distinction implies. Several ideas are brought forward. First, while home is often constructed as a female-dominated place of empowerment and security by homebirth proponents, critical readings of the domestic sphere have interpreted home as a significant site of repression and paternalism for some women (Michie 1998; McDowell 1999). Second, mothers seeking an out-of-hospital birth are drawn from several very different populations, including: young and often poor women who have a homebirth owing to socioeconomic or cultural barriers, affluent well-educated women who undertake a homebirth as part of a liberal outlook, and conservative women who have a homebirth due to patriarchal understandings of home and family (Michie 1998; Klassen 2001; Macdonald 2006). Clearly, the narratives of these groups differ significantly. Socio-economic differences are especially critical as places of birth differ widely in their cost implications. For instance, a low-income mother without insurance may feel forced into a homebirth as the cheapest option, while a more affluent mother with comprehensive health insurance may find that a hospital birth costs less out-of-pocket than a homebirth if her insurance plan does not reimburse expenses incurred outside of a hospital setting (even though the actual cost of a hospital birth is far higher). Third, new birth spaces are now opening up that arguably lead to a broader spectrum of choices over place of birth than simply home or hospital. These new spaces include birthing suites in hospitals that are supposed to provide a more relaxed "home-like" atmosphere, midwife-led units in hospitals that emphasise minimal interventions in the birth process, and birth centres that provide an institutional setting for birth that is independent of the hospital. These new settings blur the boundaries between home and hospital in different ways.

Within these discussions, birth centres are often presented as a kind of in-between place, "offering the best of both worlds" (Sharpe 1999, 94). However, they can also be framed in more radical terms, either as places of resistance to obstetric paternalism or places where the language of the homebirth community has been co-opted, thereby actually undermining homebirth (Sharpe 1999; Cheyney 2008; Worman-Ross and Mix 2013). For instance, critics of in-hospital birthing units that have been redesigned to provide a more home-like setting for birth have noted that many of the trappings of the traditional institution of the hospital have simply been hidden, rather than protocols and power relations really being questioned or renegotiated (Fannin 2004; Cheyney 2008). In this chapter, I extend such arguments to the freestanding birth centre, asking whether birth centres really provide the "best of both worlds" or instead a compromise of values in contested and power-laden circumstances?

Safety, risk and spaces of birth

The freestanding birth centre is premised on the idea that women benefit from (or are at least seeking) an alternative to the traditional medicalised hospital birth that has become the dominant paradigm in the Western world. The normalisation of the scientific approach to birth, and the idea that giving birth without easy access to technology and specialty obstetric care is risky, have led to the paradigm that the hospital is the normal site of labour and delivery in most high-income countries. This paradigm arguably has a significant influence over women's decisions around place of birth (Benoit et al. 2010). In this context, women who choose to eschew the traditional route of hospital birth have to navigate not only their own fears associated with the potential negative outcomes of an out-of-hospital birth, but also societal expectations of risk and responsible motherhood. Seeking an out-of-hospital birth therefore involves challenging forms of established authoritative knowledge, often relying instead on alternative forms of embodied and intuitive knowledge (Cheyney 2008), particularly those associated with the midwifery model of childbirth, which have a history of marginalisation from the mainstream (Newnham 2014). In this midwifery model of birth, women are framed as "naturally capable and strong, their bodies perfectly designed to carry a fetus and to give birth successfully without the high-tech surveillance and interventions of physicians in a hospital setting" (Macdonald 2006, 236). This is often seen in stark contrast to a medicalised model of birth, which pathologises the process of birth and the pregnant mother, with the accompanying expectation that the process of birth requires intervention from physicians. The decision to pursue an out-of-hospital birth is thus a highly politicised one and so choosing an out-of-hospital birth can be interpreted as an act of resistance. In this vision, women seeking an out-of-hospital birth are actively avoiding the medical gaze (Cheyney 2008) and are challenging normative medical hegemony (Worman-Ross and Mix 2013). "Homebirth practices, thus, are not simply evidence-based care strategies. They are intentionally manipulated rituals of technocratic subversion designed to reinscribe pregnant bodies and to reterritorialize childbirth spaces and authorities" (Cheyney 2011, 537).

A brief history of birth in the U.S. is instructive at this point as the move towards hospital birth was heavily influenced by the idea of risk and responsibility. This history also reveals a strongly gendered and power-laden underpinning to cultural birth norms, which can inform our understanding of current power struggles over places of birth. Traditionally, as in most of the world, birth in the U.S. was a largely female concern, with most births at the turn of the twentieth century overseen by female midwives at home. As the safety of hospitals improved, and with the development of new pain-relieving technologies, women's organisations in the early twentieth century began campaigning for women's right to hospital birth as a way to avoid risk and pain. In a more critical reading, some scholars have argued that risk was also manipulated in this era as (male) doctors associated with the newly developing field of hospital-based obstetrics began to deliberately undermine the power of competing (female) midwives by emphasising the risk of homebirths and the progress associated with hospital deliveries (Rothman 2006; Newnham 2014). By the mid-twentieth century, hospital birth was thereby becoming entrenched as the key to safe, responsible motherhood. Despite a resurgence of interest in midwifery and out-of-hospital birth in the 1960s, the dominant paradigm of the hospital as the safest place of birth had become well established, and by the turn of the twenty-first century more than 99 per cent of U.S. births occurred in hospitals (MacDorman et al. 2010).

The field of midwifery was also changed by this history of antagonism, as midwifery became sidelined and even derided. Today, traditional home-based midwives (or "lay-midwives") are in the minority, with many midwives instead operating as "nurse-midwives" as part of a hospital-based obstetric care team. Critics of this new development in midwifery argue that nurse-midwives are often forced to abandon many of the well-established patterns of holistic and comprehensive care that were great strengths of the traditional midwifery approach to birth (Bourgeault 2006).

Even today, the relative safety of different places of birth remains highly contested but is clearly a critical issue. Generally, statistical studies have found little excess risk for the *mother* associated with out-of-hospital birth, although some studies have identified a statistically significant increased risk of *neonatal* mortality or complications, particularly among first time mothers (e.g., Wax et al. 2010; Birthplace in England 2012; Cheng et al. 2013; Grünebaum et al. 2014). Some scholars suggest that Western nations that have a strong integration of homebirth with other perinatal services may show better homebirth outcomes than the U.S., where homebirth tends to be separated from mainstream maternity services (Halfdansdottir et al. 2015). Almost all studies agree that out-of-hospital deliveries are associated with lower rates of birth interventions (e.g., Birthplace in England 2012; Cheng et al. 2013; Stapleton 2013; Halfdansdottir et al. 2015). Solid conclusions on the safety of out-of-hospital birth in contrast to hospital birth are made difficult by the small numbers of women opting for homebirth, and the very small number of negative outcomes associated with birth in the affluent world, which make statistical analyses challenging. Most of our understanding has, until recently, been dependent on small observational studies where misclassified cases are often a problem (*The Lancet* 2010). Increasingly, metadata studies are being carried out to try to improve the statistical strength of analyses (e.g., Wax et al. 2010; Birthplace in England 2012). Studies of midwifery-led units, or birth centres specifically, are extremely few and far between, and often rely on even smaller numbers of cases, but generally conclude that neonatal and maternal mortality rates are similar enough to hospital settings that midwifery-led units offer a safe alternative to hospital delivery for women at low risk of birth complications (e.g., Birthplace in England 2012; Stapleton 2013).

The risk associated with an out-of-hospital delivery is therefore extremely hard to quantify and is still debated, leaving experts and mothers alike often unsure of how to proceed. In addition to the challenges of statistical assessments of risk, there is also considerable disagreement over the importance of different types of risk. Obstetricians' organisations have tended to focus on neonatal and maternal injury and death as the pre-eminent causes for concern, while those following a midwifery model of care have also given consideration to the potential damage caused by "unnecessary" interventions in the birth process and the importance of maternal choice over place of birth, while also acknowledging the pivotal importance of physical outcomes. Owing partly to these different under-standings of relative risk, midwifery colleges in Australia, the U.S., and the U.K. (among other countries) have leant their support to homebirth, while obstetric organisations in Australia and the U.S. have not traditionally sup-ported homebirth (Newnham 2014; Licqurish and Evans 2016). Of relevance to this chapter, the American College of Obstetricians and Gynecologists (ACOG) has long argued that hospital is the safest place for birth and has strongly dis-couraged homebirth; however, ACOG has recently endorsed freestanding birth centres as a safe place for labour and delivery for women with low risk preg-nancies (ACOG 2011). Thus, the birth centre has now become a professionally sanctioned "safe" space for birth in the U.S.

All these debates over risk are critical to my argument here as the birth centre has been positioned as potentially operating as an in-between space between hospital and home, and this can be argued also to be the case in the notion of risk. As women negotiate the birth landscape, Craven (2005) notes that women can feel the need to reclaim "respectable motherhood" from medical officials who challenge alternative approaches to birth, with mothers potentially facing harsh criticism for putting their own needs above their baby's if they opt for an out-of-hospital birth. In the discussion that follows, I therefore consider whether a major role of the birth centre is in providing a way for women to reclaim "respectable motherhood" while seeking the birth experience of their choice.

Methods

The results reported here are drawn from a study of out-of-hospital births in the Twin Cities, Minnesota, with data collected from 2011 to 2013. The focus of this part of the study is interviews with 24 mothers who had intended an out-of-hospital birth within the past two years, but the broader study also included interviews with practitioners involved in out-of-hospital birth and partners of birth mothers.

The rate of homebirth in Minnesota was only 0.72 per cent in 2009, placing it twentieth among the U.S. states (*Star Tribune* 2012). Nonetheless, Minnesota has experienced a resurgence of interest in out-of-hospital birth over the past 20 years, seeing an increase of 25 per cent between 2003/4 and 2005/6 and a further increase by 2009 that pushed its rate just above the national average (MacDorman et al. 2010; Star Tribune 2012). The Twin Cities (Minneapolis and St. Paul) are widely recognised in the U.S. as politically progressive and are host to a variety of groups supporting alternative birth. Lay midwifery is legal and regulated in Minnesota and legislation from 2010 licenses and regulates freestanding birth centres. These state-based regulations have been important in encouraging insurance companies to reimburse out-of-hospital births; an increasing number of insurance companies are now covering birth centre births and a few are covering homebirths. As of April 2014, five birth centres had been accredited and were operating in the state: three in the Twin Cities, one in Duluth, and one in rural Western Minnesota, primarily serving women from Amish and Mennonite communities. The Twin Cities stands as a regional centre for out-of-hospital birth and the results reported here should not be considered representative of U.S. women or even Minnesotan women. Instead, my goal is to explore individual women's motivations for out-of-hospital birth in order to build up an understanding of the factors that are leading to the current resurgence of interest in birth centres.

Mothers were recruited via advertisements in local mothers' and midwives' listservs and a birth centre newsletter and were invited to contact the researcher if they would like to participate in the study. More than 30 eligible women quickly contacted the researcher, enthusiastic to take part—a clear demonstration of the great interest in sharing birth stories within this community. It is important to note that mothers who were not keen to share their birth stories, perhaps owing to negative birth outcomes, are unlikely to be captured in this surveying method—a clear limitation of the study. Initially, all eligible participants were interviewed as they made contact, but after about ten interviews had been completed further interviewees were purposely selected to include women who might provide different viewpoints. Contacts in the birth community provided assistance in identifying individuals with a diversity of birth stories. It was notable that my respondents were heavily drawn from among mothers undertaking an out-of-hospital delivery as part of a liberal outlook; only one respondent could be considered to potentially represent a religiously conservative motivation for homebirth, and no respondents fell into the category of young/poor/poorly educated and having an unintentional homebirth. Having said this, respondents were far more diverse in terms of their income, level of education and profession than expected, with some women reporting that they relied on government food assistance, others that they were professionally trained as lawyers and nurses—the affluent/professional/liberal homebirth mother is perhaps more stereotype than reality. There was also a notable lack of ethnic diversity among respondents (all were white); this was considered acceptable given that homebirth mothers in the U.S., and particularly in Minnesota, are pre-dominantly white. What did unite all respondents was a very high level of self-taught knowledge of birth, even if their formal education was limited. In summary, it is important to reiterate that the respondents in the study were not selected to be representative of the overall population of mothers seeking an out-of-hospital birth in the U.S., or even in the Twin Cities. The study was designed to begin to explore issues of importance to the out-of-hospital birth population, rather than to draw statistical conclusions about characteristics of that population.

In-depth interviewing was essential to exploring the particularities of experience that this approach is trying to identify. Interviews were conducted in coffee shops or interviewees' homes, according to interviewee preference, and lasted about an hour. Interviewees were asked to sign a consent form, consistent with IRB (institutional review board) protocol, which included a request for permission to tape-record the interview. The goal was to identify multiple "case studies" that explored the reasons for seeking an out-of-hospital birth experience. As Yin (2003) notes, comparing multiple case studies in this way is an ideal starting point for exploring phenomena where context is believed to be extremely significant to the topic of study. This case study approach begins from the premise that in-depth and extensive methods of enquiry are essential to identifying the diverse variables that are potentially of interest in explaining a particular phenomenon. Gerring (2007) suggests that as more case studies are collected, at some point case study research becomes "cross-case research", where the large number of case studies allows the researcher to begin to identify themes of relevance to the whole population. Generating themes of *relevance* to the broader population was the goal of this research, while acknowledging that the ideas reported will not be *representative* of that population.

In order to analyse my results in a way that might uncover some of these broader themes, I used a thematic networks approach (Attride-Stirling 2001). First, I pulled *basic themes* from the text. These are statements of belief that relate to the phenomenon under study, in this case motivations for an out-of-hospital birth. These belief statements were then analysed in order to generate *organising themes*; for instance, the notions of "power" or "risk". Finally, the organising themes were grouped into *global themes*: groupings of ideas that, as a whole, develop an argument about an issue (Attride-Stirling 2001). In this case, global themes were identified that elucidate the perceived value of a birth centre birth.

The results reported here consider women's narratives around home, hos-pital and birth centre, teasing apart the significance of these different sites of birth and what we can then conclude about the role that birth centres are currently playing on the birth landscape. The results come from 24 women, who had intended to deliver a baby in an out-of-hospital setting within the past two years (*intention* is key here as women may not end up giving birth where they had planned). Among my respondents, 14 had intended a home-birth and ten a birth centre delivery. Two women who intended a birth centre delivery had an unplanned homebirth owing to the rapidity of their labour. Three of the intended homebirths transferred to hospital during labour because of complications (stalled labour or breech birth). In addition, two women had *unassisted* homebirths: one planned and one unplanned. A total of 11 women referred only to the birth of their first child in their responses; 13 respondents also referred to previous births in their interviews, adding a further 11 hospital births, six homebirths, and one birth centre delivery of older siblings. All respondents are referred to by pseudonym throughout the forth-coming discussion. The notation "BC" is used to identify women whose most recent birth was intended to occur at a birth centre and "HB" for respondents who had intended a homebirth.

The appeal of out-of-hospital birth

In order to understand the significance of choosing a birth centre birth, we first have to consider the attraction of an out-of-hospital birth to expectant mothers. As I report elsewhere (Hazen 2017), my interviewees discussed five main themes in explaining the attraction of an out-of-hospital birth (Table 7.1). First, most respondents seeking an out-of-hospital birth did so because they sought to minimise interventions in the birth process. Second, and related to this, was the desire to receive care from a midwife, who was seen as far less likely to intervene unnecessarily in the birth process than an obstetrician. Midwives were described as seeing birth as a normal, natural process, which was expected to be successful with minimal intervention. By contrast obstetricians were reported by many of my respondents to be trained to expect complications, leading to higher rates of interventions, particularly caesarean sections. Although midwives can operate in hospital or out-of-hospital settings, for many women it was the combination of mid-wifery care and an out-of-hospital setting that seemed most likely to provide the low-intervention birth that they sought. Third, hospital was seen as a place of disempowerment for expectant mothers, where birth options are

Table 7.1 Key themes related to the attraction of out-of-hospital birth

THEME	HOME	HOSPITAL
Interventions	Place of few interventions	Place of many interventions
Models of care	Associated with midwifery	Associated with biomedicine
Empowerment	Place of empowerment	Place of disempowerment
Retreat	Place of retreat	Place of disturbance
Philosophy	Associated with holistic philosophy of birth and "natural" lifestyle	Associated with the pathologising of birth

restricted by stringent protocols. For many women, getting out of the hospital was seen as the first step in taking charge of their birth. Fourth, home was seen as a place of retreat where the mother could be relaxed, thereby facilitating the process of labour. Finally, many women described how out-of-hospital birth was part of a wider philosophy of letting nature take its course, related to lifestyles of environmental and social responsibility. Similar findings have been presented in studies of Finland, Sweden and Canada (Hildingsson et al. 2003; Kornelsen 2005; Jouhki 2012; Murray-Davis et al. 2012; see also Whitson in this volume), suggesting some consensus among women in high-income contexts over their motivations for seeking an out-of-hospital birth.

I will not reiterate these arguments further here, but instead want to expand on how the physical and psycho-social spaces of alternative places of birth are understood by women seeking an out-of-hospital delivery, in order to begin to understand the role of birth centres specifically in meeting these needs. Is the birth centre an in-between space where women can seek the best of both worlds, or is it instead a compromise between the birth that respondents would like and the birth that they and others feel they ought to have? This is a key question: in the context of discussions around risk and responsible motherhood noted above, reframing the birth centre as a possible compromise suggests that mothers may be negotiating complex pressures from medical practitioners, partners, family members and broader society in undertaking a birth centre birth.

Attractions of the birth centre

Providing a "homelike" environment

A dominant theme among my respondents was the importance of feeling comfortable in the place chosen for labour and delivery. Many women mentioned how the small intimate environment of the birth centre was a place where they could feel comfortable and know that they were known and respected by members of staff. The following comments were typical:

> Everybody was just so nice. And yes, my OB [obstetrician] with my older daughter was nice too. But just the way that they talk to you [at the birth centre] and…. it was just such a smaller atmosphere, it didn't feel like I'm coming into this factory and that I'm just a pawn.
>
> (Hannah BC)

> I mean, you know, even the office manager knows us by name and face. I liked the small community feel to it.
>
> (Rosalie BC)

Rosalie went on to note that this intimate atmosphere had a tangible impact on her, by suggesting that her success in breastfeeding was in large part due to the personalised care that she received from her birth centre midwives.

Several women contrasted these personalised experiences at a birth centre with previous experiences of hospital-based perinatal care. Clara's (BC) story, although more extreme than most, is indicative of the stark contrasts that were drawn between birth centre and hospital care by many respondents:

> I'd had one consultation with [a local birth centre] and I was in the system at [a local hospital]. When the miscarriage came I had a stark comparison between the way the two systems handled it. My experience with [the hospital] was calling them and being totally scared and bleeding...and being told, "I'm so sorry, we're between shifts. Can we call you back?" ...And so, I waited twenty minutes, bleeding, scared, my husband was gone. And they called me back twenty minutes later and then the lady was like "Oh, I'm just calling up the computer. Bear with me, [Sarah]!" And I'm like, "I'm not [Sarah]—I'm [Clara]!" ...And weeks later they were calling me—after I had lost the baby—calling asking if I wanted an amniocentesis. And I was like, "No, really. I just lost the baby and I don't want an amniocentesis." And it was horrible. But since I had already had the one consultation with [the birth centre], I had a relationship with the midwife there already. And she was somebody that I was able to call. And she got me through the whole thing. She was really, really supportive and everything she said made a whole lot of sense.

Other respondents noted ways in which their birth centre attempted to generate an explicitly *home*like environment. Melissa (BC), for instance, noted that she and her partner were encouraged to bring a meal for after the birth and described how her husband had heated up enchiladas in the birth centre kitchen during the immediate post-partum period. Beth (BC) noted that her birth centre was very flexible about who could be in the birthing room, so that she could pick and choose the people there. Indeed, for some women, their decision over which birth centre to choose was explicitly framed around which one made them feel most at home, as explained by Amy (BC):

> When I was touring both birth centres, the first one had kind of a clinical feel to it, a bit like a doctor's office. It was at the end of a strip mall. The rooms were nice, and the staff were nice enough, but it just seemed like I would have been just another patient. The place in Minnesota, as soon as I walked in the door, it just felt like walking into a relative's house. It wasn't my house, but it just felt very homelike and welcoming and comforting. And all the staff was just professional and friendly. It just all clicked. It was a gut feeling more than anything—it just felt right.

This idea that the birth centre should not have a clinical feel to it was common and seemed to have been factored in to the very physical design of many birth centres. Amy (BC), for instance, noted that at her birth centre the examination rooms for prenatal appointments are kept separate from the delivery rooms so that the clinical operations of the birth centre do not interfere with your birth experience. Many women mentioned the décor of the room where they had chosen to give birth, noting how it helped to provide a pleasant atmosphere for the birth. Amy (BC), for instance, stated:

> They have three rooms to choose from and you rank them. You know, this is the first choice. If that room is taken when I go into labour, here is my second choice. I was able to have my first-choice room which is decorated in chocolate brown and light blue, which are the colors we had chosen for my baby's room. It has a queen-sized bed, with all the bedding, pillows…and the tub was in the room with the bed…just kind of a Jacuzzi-sized tub. And dim lighting, which could be turned up or down as needed. Then they had a large bathroom with a sink, toilet, a shower with a bench in it.

None of my respondents suggested that these trappings of décor actually provided an explicitly home-like setting. Instead, most women described the setting as luxurious, often noting with mirth that this is in contrast to their own home. The commonest analogy used was that the birth centre resembled a bed-and-breakfast. Melissa's (BC) description of her birthing suite bears repeating in full:

> My view during labour was of a glass chandelier and I just adore chandeliers. It's just exquisite. The woodwork alone in the reception area is gorgeous. But then the three birthing rooms, and they let you have your choice of birthing room so that you can visualise that (assuming no other mom is using that at the time). But, you know, you walk in and it's this huge queen bed, huge Jacuzzi, handicapped-accessible bathroom, you know, real towels in the bathroom. Color—it's not just white. It's up-to-date. You know, it looks like an exquisite bed-and-breakfast…And they have a beautiful garden that you can walk in. Just lovely. They use a mister-humidifier with essential oils in it. It just smells delicious. It looks delicious. And even though they have to have linoleum floor to meet health code, it looks like wood. Oh, and my bed had black satin sheets— did I mention that? I picked the white-and-black room. I like that combo—real sheets!

This luxurious experience at the birth centre was something that some of my respondents had explicitly sought. For some women, this involved a considerable financial commitment, as Clara (BC) noted:

So, we're only going to have one kid probably. So, it was a once in a lifetime experience and we wanted to do this absolutely right. We wanted the best thing ever and so we splurged. We knew that we were probably going to have to pay full price. But we saved for it. Fortunately, we started saving when we started trying to conceive and so we saved way more than we meant to because it took so long! [laughter]

For other women, however, this luxury was more than they could afford. At an average of around $6,000–$8,000 for birth, pre-natal and post-natal care (as reported by my respondents), a birth centre birth is relatively expensive in contrast to a hospital delivery when it is largely covered by insurance or a homebirth (usually under $4,000 for comprehensive peri-natal care, according to my respondents). At least one of my respondents had wanted to have a birth centre birth but finally opted for a homebirth as the cheaper option. At the time of interviewing, only a handful of insurance plans would cover a birth centre birth and the financial implications of a birth centre birth diverged significantly among my respondents. A couple of women were lucky enough to have insurance cover their birth centre birth, but most ended up paying out of pocket—often after lengthy negotiations with their insurance provider. Thus, the birth centre delivery was, for most women in this study, a financial commitment above and beyond other options, often to the tune of several thousand dollars. (It is important to note that this situation has evolved rapidly over the past five years, with a larger number of insurance companies now covering the cost of a birth centre delivery, thereby leveling the playing field somewhat.)

Beyond the physical surroundings, another major attraction of the birth centre was the flexibility it offered. Several respondents noted that they approached the birth centre specifically to have the birth that they wanted, as explained by Beth (BC):

> I met with the midwives at [the birth centre] and they kind of laughed as I was going through all of the things that I didn't want to do. They were like, "well we don't do that!" we don't even consider doing most of the things I was listing. I was going to at least have a shot at the birth experience I wanted if there would be people there to support me.
>
> (Beth BC)

Other respondents noted the time and support that midwives devoted to each patient, often suggesting that appointments would last over an hour and cover a wide variety of topics, including family life, nutrition and other holistic aspects of pregnancy and general wellbeing. Even Beth's enthusiastic endorsement of her birth centre's philosophy came with a caveat, however:

> I think, ideally, I would have got a little bit more direction from my care provider. The midwives were very hands-off. They wanted me to have the

birth experience that I wanted to have so they just wanted to stay out of the way, which I completely understand and respect, but as a first-time mom not knowing what to expect, just having a little more direction and hand-holding through the process would have made me personally feel better.

How much intervention and guidance mothers want in their birth is clearly specific to the individual. For those actively seeking a low-intervention birth, however, an out-of-hospital experience was often seen as critical.

Birth centre as separation

For some women, it was not just that the birth centre was an *attractive* location for giving birth, it was also significant that it was *separate* from the home. For Claire (BC) this feeling was particularly strong, although hard for her to articulate:

I did like the separation of the birth space from the house, like afterwards when we came back, I did like that separation. Just to have that... a place where I could go to for all my visits, you know, just to have that place, separate.

When pushed to explain this idea further, Claire went on to clarify, "It's almost like a sacred space, you know like going to church for worship or whatever."

For other women, the desire to have their baby in a place separate from their home was more tangible. Several women noted that having the birth in their own home would have left them uncertain of what to do with their other children. Others noted that having a homebirth involves feeling that you have to *host* the birth. Cara (HB), for instance, outlined the anxiety that went into preparing the house for her homebirth:

I was surprised at the need to prepare the house to have a baby involved with my second homebirth. Just the amount of energy extended in thinking about taking care of the midwives' food needs. Not that they really have any, but I felt a lot of pressures with another layer of nesting to prepare the house for a birth ... You know, thinking about having all of your supplies together and everything in the right place. Having clean sheets on the bed. Basic hospitality.

The idea of having a birth centre birth appealed to some respondents precisely because they did not have to go through these preparations. Similarly, many respondents mentioned, often laughing, that they really did not want to clean up the mess of birth in their own home. In such cases, going to a birth centre removed the psychological and practical mess and disruption of birth from the home.

Navigating risk

A final attraction of the birth centre was, as expected, in navigating risk. For some women, the birth centre acted as reassurance that they were in an institutionalised setting with trained medical professionals in charge. In this context, the birth centre is seen as a reasonable middle ground between the *risky* setting of home and the *security* of hospital. Hannah's (BC) comments are typical of this sub-set of women: "I didn't want to do a homebirth because I was scared of it, but once I saw that there were birth centres I felt like: this is OK." Similarly, Beth (BC) felt reassured that the birth centre would have equipment and expertise on hand in case of emergency:

> I was still kind of feeling that I was on the older end of being pregnant for the first time, and with my family's medical history there was a chance that I would develop preeclampsia. So, I just wanted the security of feeling that there was more help available to me if I needed it.

For others, the belief that birth centre staff would facilitate a transfer to a hospital, should the need arise, was a major attraction.

A separate subset of women reflected that this extra feeling of security was probably little more than a feeling, noting that there is little available at a birth centre that you could not have at home during a birth. Cara (HB), who intended a birth centre birth for her first child (although ended up having a homebirth owing to a very rapid labour), aptly summarises the feelings of this second sub-group of my respondents:

> I think that there's a perception that they have something at a birth centre that you wouldn't get at a homebirth. Maybe it's an idea that they'd have something for an emergency. But, at the end of the day, you know, homebirth midwives have their oxygen with them, they have suture kits with them, they have oxytocin or Pitocin with them.

In spite of this, many women were nonetheless reassured by the idea that there were protocols in place that would offer extra security. Hannah (BC) noted:

> Well, you're still in *their* location. So, they have like all these back-up plans. And I know they have them at home too. But it felt a little bit easier also to tell family. You know: we're still going to this centre. You tell them the homebirth and they're like: "What's wrong with you?" You could still get the midwife care but you're still at a centre where... just being in an institution still had that comfort feeling behind it.

Hannah's statement is telling in a number of ways. Not only does she emphasise the potential security (real or imagined) that she received from the institution of the birth centre, but she also reveals that part of the value of this security is really for those around her, rather than for her personally. Several women noted the emotional stress of electing to have an out-of-hospital birth, owing to real or perceived censure from family members or the broader community. For instance, Lacey (BC) stated:

> I had been pregnant previously and I know my mum she kind of freaked out at a homebirth. But the birthing centre, after I told her all the pros and cons of it and that it was right across the street from the hospital, you know she calmed down and she was fine with it.

Claire (BC) similarly noted that having her baby in a birth centre was more acceptable to her family than a homebirth: "When I said 'birthing centre' and I explained that this was a place that midwives work at and these people are trained at birth, they were understanding and felt reassured that there was a hospital nearby." Claire (BC) noted that her fiancé, too, had been reassured by the idea of a birth centre, leading them to compromise on a birth centre birth: "I first wanted to do a homebirth, but my fiancé was worried about the potential risk so a birthing centre was a nice medium." For other women, it was rather societal censure that was the concern. Cara (HB) notes, admittedly in jest but underscoring a more serious point, "I think going somewhere just feels more like going to the hospital. And then you don't have to tell people you've had a homebirth! [laughter]"

Discussion: Birth centres as compromise or the best of both worlds?

Birth centres offer different things to different people, but several common threads can be identified from the results reported here. First, many women were attracted to the philosophical underpinnings of a midwife-led delivery. In most cases this commitment to a minimally-interventionist birth could be achieved equally well in either a birth centre or home setting. For some women electing to have a birth centre delivery over a homebirth, the extra value came from the separation of home and birthing space. Sometimes, this separation was important for tangible reasons—keeping the home space "clean" of the mess of birth or finding a relaxed and pleasant space to deliver away from the chaos and everyday responsibilities of the home. In other cases, this separation was important for less tangible reasons—the need for a "sacred space" to give birth, or the reassurance received by handing over responsibility for the birth to an institutionalised setting with its own set of protocols.

Perhaps more importantly, however, the idea of risk was central to most mothers' discussions around place of birth, with a sizable proportion of my respondents explicitly mentioning that they chose a birth centre over a homebirth because they felt more comfortable with their decision from a risk management perspective. Although several women acknowledged that this extra sense of security may be more perceived than real, the *perceived* safety of the birth centre setting was nonetheless valuable for many women through reassuring family members and avoiding social censure over their desire for an out-of-hospital birth. All my respondents were united in their desire to have a safe birth but being seen to make *responsible* choices (Craven 2005) was also of importance to many, especially as many respondents were bothered by nagging doubts about the implications of their personal decisions on their unborn child. These ideas resonate with broader discussions of "responsible motherhood". For many years, scholars have interpreted reproduction-related activities as sites of control, arguing that the medical establishment and broader society have frequently tried to maintain control over women by emphasising the need for expert knowledge in navigating what were once everyday activities (e.g., birth, breastfeeding and childrearing), often utilising Foucauldian understandings of knowledge/power as key to achieving this (see, for example, Hays 1996; Wall 2001; Newnham 2014). In such analyses, the mother is viewed as restricted and marginalised as a subject with her own needs and wants; instead her actions are interpreted only in the context of their influence on her child. Wall (2001, 604), for instance, explains in the context of breastfeeding: "Mothers and mothers' needs disappear from view here and mothers' behavior becomes legitimately subject to public scrutiny and moral authority. Women become, in part, builders of better babies or burdens on the social safety net." I argue that decisions over place of birth are equally subject to this framework of responsible motherhood, with mothers' wants and needs potentially marginalised in society's desire to emphasise the need for a "safe" delivery of future generations. In so doing, the structural constraints that might influence a woman's reproductive choices are minimised, while a moralising discourse of "responsible motherhood" is emphasised instead.

For some women, birth centres were explicitly framed as the "best of both worlds" within these debates: a place where they could get the minimally invasive birth that they desired (perhaps even with a little bit of luxury thrown in), but with the key benefit of also being able to ensure that they are viewed as "responsible mothers" birthing their babies in the safety of the institutionalised space of the birth centre. As Lacey (BC) notes:

> For me it seemed like the best of both worlds. Because it was like a home environment, it was very calm. It had a beautiful, wonderful tub to sit in, which was awesome. And it was a very calm environment—it had the midwives and everyone was there. Everything that they needed was there. And to calm my mum's mind, it did have the hospital right across the street, even though I knew that we probably would never need it.

In more critical terms, however, the birth centre can instead be viewed as a place of compromise. Several women noted explicitly that the birth centre had been a compromise between themselves and another family member (often their partner, mother or mother-in-law), or reflected that using a birth centre to some degree bowed to social pressure for "a place to go" to birth their baby. Annie (BC) notes this compromise in no uncertain terms: "John [the baby's father] was OK with the compromise: I wanted a homebirth, John wanted a hospital birth, so we compromised". Similarly, Cara (HB), referring to her intention for a birth centre birth for her first child, notes:

> The idea of a birth centre was less over-the-top than a homebirth. And so that was the closest homebirth option at the time and was ... I mean my husband was more hesitant about an out-of-hospital birth and was more comfortable because it was somewhere to go. Which, at the time I knew was kind of a farce.

Cara's true feelings about the situation become evident as she notes that the situation was "kind of a farce". Deep down she does not fully buy in to the idea that the birth centre is necessarily a safer option, but for her it worked out as the most expedient choice at the time, given the social context she found herself in. In such cases, the birth centre becomes a place that gets women nearer to the birth experience that they really want, but in a socially acceptable way that helps stave off nagging doubts about the safety of their decision by demonstrating to family and broader society that they are behaving as responsible citizens.

Conclusion

Birth centres have often been positioned in academic literature and in the promotional literature of the birth centre industry as providing the "best of both worlds". In this reading, the birth centre offers the benefits of the homebirth experience through a relaxed home-like environment and the minimally invasive care of a midwife, and yet meets concerns over safety by providing an institutionalised setting with considerable expertise for a safe and successful delivery. A closer look at the experiences of women seeking an out-of-hospital delivery suggests that the reality is more complex.

While many women appreciated the comfort and luxury of the birth centre, few respondents suggested that it was anything like home. For some it was "more than home"—akin to a stay in a luxurious bed-and-breakfast, while for others it was acknowledged as being just another institutionalised setting—albeit one with a more acceptable philosophical approach to birth than the hospital.

The idea that the birth centre offered additional safety was widely critiqued among my respondents. Although many interviewees noted that they had initially been drawn to the potential additional safety of a birth centre birth over a homebirth, many went on to qualify this by suggesting that they were not sure if this was entirely true, even if they continued to get some reassurance from the idea. For many respondents, the true value of the idea that a birth centre might be safer was through making the most of the perception among the general public or their family members that birth centres are safer than homebirths. Remember that even the American College of Obstetricians and Gynecologists (ACOG) now endorses birth centres as a safe alternative to hospital. By allowing a family member to consider the birth centre as a "safe option", respondents were able to conform to pressures to be "responsible mothers" while still pursuing the birth that they wanted. Even those who did not receive censure from family members over their decision for an out-of-hospital birth often mentioned in passing that a birth centre birth raised fewer eyebrows than a homebirth, as broader society still seems more comfortable with the idea of having a place to go for labour and birth.

Given the multiple and varied perspectives reported here, the argument that birth centres provide the "best of both worlds" does not seem to offer the whole story. While some women reported that their birth centre experience had been exactly what they sought—usually justifying this position by reporting that they got the comfort of home with the extra security of going to a special place for birth—many women problematised this idea. For some, the birth centre represented a compromise between themselves and a more risk-averse family member, for others the birth centre represented a more socially acceptable way to get closer to the style of birth they wanted, allowing them to conform to societal constructions of "responsible motherhood". For all women, it was clear that power structures at the household and broader societal scale influenced their decisions around place of birth, leaving the decision over where to give birth as a frequent site of compromise.

References

ACOG [American College of Obstetricians and Gynecologists]. 2011. Committee Opinion on Planned Homebirth. Available at: https://www.acog.org/Resources_And_Publications/Committee_Opinions/Committee_on_Obstetric_Practice/Planned_Home_Birth. Accessed 5. 28. 14.

Attride-Stirling, J. 2001. "Thematic Networks: An Analytic Tool for Qualitative Research." *Qualitative Research* 1: 385–405.

Birthplace in England. 2012. Birthplace in England – New Evidence. *SDO Network Research Digest*, issue 3. Available at: http://nhsconfed.org/~/media/Confederation/Files/Publications/Documents/birthplace-england_130612.pdf

Benoit, C., Zadoroznyj, M., Hallgrimsdottir, H., Treloar, A. and Taylor, K. 2010. "Medical Dominance and Neoliberalisation in Maternal Care Provision: The Evidence from Canada and Australia." *Social Science & Medicine* 71(3): 475–481.

Bourgeault, I. 2006. *Push! The Struggle for Midwifery in Ontario.* Quebec City, Canada: McGill-Queen's University Press.

Cheng, Y., Snowden, J., King, T., and Caughey, A. 2013. "Selected Perinatal Outcomes Associated with Planned Home Births in the United States." *American Journal of Obstetrics and Gynecology* 209: 325 e1–8.

Cheyney, M. 2008. "Homebirth as Systems-challenging Praxis: Knowledge, Power, and Intimacy in the Birthplace." *Qualitative Health Research* 18(2): 254–267.

Cheyney, M. 2011. "Reinscribing the Birthing Body." *Medical Anthropology Quarterly* 25(4): 519–542.

Craven, C. 2005. "Claiming Respectable American Motherhood: Homebirth Mothers, Medical Officials, and the State." *Medical Anthropology Quarterly* 19(2): 194–215.

Fannin, M. 2004. "Domesticating Birth in the Hospital: Family-centered Birth and the Emergence of Homelike Birthing Rooms." In *Life's Work: Geographies of Social Reproduction*. K. Mitchell, S. Marston, and C. Katz, eds. 513–535. Malden MA: Blackwell Publishing.

Gerring, J. 2007. *Case Study Research: Principles and Practices*. N.Y.: Cambridge University Press.

Grünebaum, A., Mccullough, L., Sapra, K., Brent, R., Levene, M., Arabin, B., and Chervenak, F. 2014. "Early and Total Neonatal Mortality in Relation to Birth Setting in the United States, 2006–2009." *American Journal of Obstetrics and Gynecology* 211 (4): 390.e1–390.e7.

Halfdansdottir, B., Smarason, A., Olafsdottir, O., Hildingsson, I., and Sveinsdottir, H. 2015. "Outcome of Planned Home and Hospital Births among Low-risk Women in Iceland in 2005–2009: A Retrospective Cohort Study." *Birth* 42(1): 16–26.

Hays, S. 1996. *The Cultural Contradictions of Motherhood*. New Haven: Yale Press.

Hazen, H. 2017. "'The First Intervention is Leaving Home': Reasons for Electing an Out-of-hospital Birth among Minnesotan Mothers." *Medical Anthropology Quarterly* 31(4): 555–571.

Hildingsson, I., Waldenstrom, U., and Radestad, I. 2003. "Swedish Women's Interest in Home Birth and In-hospital Birth Center Care." *Birth* 30(1): 11–22.

Jordan, B. 1993. *Birth in Four Cultures: A Crosscultural Investigation of Childbirth in Yucatan, Holland, Sweden, and the United States*. Long Grove, IL: Waveland Press, 4th ed.

Jouhki, M. 2012. "Choosing Home Birth—The Women's Perspective." *Women and Birth* 25: e56–e61.

Klassen, P. 2001. "Sacred Maternities and Postbiomedical Bodies: Religion and Nature in Contemporary Home Birth." *Signs* 26: 775–810.

Kornelson, J. 2005. "Essences and Imperatives: An Investigation of Technology in Childbirth." *Social Science and Medicine* 61(7): 1495–1504.

The Lancet. 2010. "Home Birth—Proceed with Caution." *The Lancet* 376 (July 31).

Laws, P., Lim, C., Tracy, S., and Sullivan, E. 2009. "Characteristics and Practices of Birth Centres in Australia." *Australian and New Zealand Journal of Obstetrics and Gynaecology* 49: 290–295.

Licqurish, S., and Evans, A. 2016. "'Risk or Right': A Discourse Analysis of Midwifery and Obstetric Colleges' Homebirth Position Statements." *Nursing Inquiry* 23 (1): 86–94.

Longhurst, R. 2008. "At Home with Birth." In: *Maternities: Gender, Bodies and Spaces*. NY: Routledge, 81–100.

Macdonald, M. 2006. "Gender Expectations: Natural Bodies and Natural Births in the New Midwifery in Canada." *Medical Anthropology Quarterly* 20(2): 235–256.

MacDorman, M., Menacker, F., and Declercq, E. 2010. "Trends and Characteristics of Home and Other Out-of-hospital Births in the United States, 1990–2006." *National Vital Statistics Reports*, 58, no. 11 (2/3/2010).

McDowell, L. 1999. "Home, Place and Identity." In: *Gender, Identity, and Place: Understanding Feminist Geographies*. Minneapolis, MN: University of Minnesota Press, 71–95.

Michie, H. 1998. "Confinements: The Domestic in the Discourse of Upper Middle Class Pregnancy." In: Aiken, S. et al. *Making Worlds: Gender, Metaphor, Materiality.* Tucson, AZ: University of Arizona Press, 258–273.

Murray-Davis, B., McNiven, P., McDonald, H., Malott, A., Elarar, L., and Hutton, E. 2012. "Why Home Birth? A Qualitative Study Exploring Women's Decision-making about Place of Birth in Two Canadian Provinces." *Midwifery* 28: 576–581.

Newnham, E. 2014. "Birth Control: Power/knowledge in the Politics of Birth." *Health Sociology Review*, 23(3): 254–268.

Rothman, B. K. 2006. "Laboring Then". In *Laboring on: Birth in transition in the United States.* W. Simonds, B. K. Rothman and B. Meltzer Norman, eds. New York: Routledge.

Sharpe, S. 1999. "Bodily Speaking: Spaces and Experiences of Childbirth." In: *Embodied Geographies: Space, Bodies and Rites of Passage.* E. Teather, ed. 91–101. NY: Routledge.

Stapleton, S., Osborne, C., and Illuzzi, J. 2013. "Outcomes of Care in Birth Centers: Demonstration of a Durable Model." *Journal of Midwifery and Women's Health* 58 (1): 3–14.

Star Tribune. 2012. "Home Births a Rising Trend in Minnesota." January 26.

Wall, G. 2001. "Moral Constructions of Motherhood in Breastfeeding Discourse." *Gender and Society* 15(4): 592–610.

Wax, J., Lucas, F., LamontM., et al. 2010. "Maternal and Newborn Outcomes in Planned Home Birth vs Planned Hospital Births: A Meta Analysis." *American Journal of Obstetrics and Gynecology* 203(243): e1–8.

Worman-Ross, K. and Mix, T., 2013. "'I Wanted Empowerment, Healing, and Respect': Homebirth as Challenge to Medical Hegemony." *Sociological Spectrum* 33: 453–481.

Yin, R. 2003. *Case Study Research: Design and Methods* (3rd ed). London: Sage Publications.

8 "My germs, my space, my stuff, my smells"

Homebirth as a site of spatialised resistance in Appalachian Ohio

Risa Whitson

In the 1800s and early 1900s, almost all women in the United States gave birth at home with the assistance of a midwife. However, as the speciality of medical obstetrics became more established in the early twentieth century, and with the concurrent regulation and—in some places illegalisation—of midwifery, the percentage of hospital-based births increased rapidly. By 1950, 88 per cent of all births in the U.S. took place in the hospital and by 1969 hospital birth had become ubiquitous, with rates of over 99 per cent (Boucher et al. 2009). Today planned homebirth in the United States remains very rare, currently at just below 0.9 per cent of all births. In spite of this small percentage, planned homebirth is at its highest rate since the early 1950s, when hospital birth became normative (MacDorman et al. 2014). Furthermore, despite strong institutional pressure against it—including from the medical establishment, insurance companies, care providers and society as a whole— the rate of homebirth has gone up slightly but steadily over the last ten years, from 0.56 per cent in 2004 to 0.89 per cent in 2012 (Betrán et al. 2016). This slow, but steady, increase in homebirths is one aspect of a larger movement which seeks to create alternatives to birth as an institutionalised and medicalised process, including through unmedicated birth, birthing centres and the growth of certified nurse midwives in obstetric practice. However, for families that choose planned homebirths, place is critical to the alternative understanding of birth that they envision and enact.

In this chapter, I consider the geography of birthing at home by asking in what way place informs the meaning of homebirth for mothers. While many women challenge the norms of medicalised birth through resisting medical intervention, planning unmedicated births or choosing to birth in hospitals and birth centres with women- and child-friendly birth practices, homebirth spatialises this resistance in a unique way. This chapter draws on 13 in-depth interviews with women who birthed at home to analyse homebirth as an act of explicitly spatialised resistance to biomedical norms and institutional control at two interconnected scales: the body and the home. In order to avoid problematic assumptions that overlook the complex nature of consent and control in the context of reproductive decisions, unlike other literature on homebirth, this chapter does not focus on reasons why women choose to birth at home. Rather, I examine how the place of the home is a critical component in the way that women *understand* and *narrate* their birth experiences. Specifically, I argue that women describe birthing at home as enabling them to: normalise birth, support their mobility and bodily autonomy during birth, permit them increased control over space during birth and re-spatialise the experience of birth. This chapter thus seeks to contribute to an explicit spatialisation of the research on birth by exploring the way that home functions in women's narratives to locate homebirth as a site of resistance to norms of institutionalisation and medicalisation.

Homebirth in context

Academic research on homebirth is framed largely through the lenses of medicine and nursing, with a primary focus on analysing the associated risks and outcomes, often in comparison to hospital birth, but also in relation to free-standing birth centres (Jonge et al. 2013; Wilbur et al. 2015). In spite of extensive research in this area, scholars remain inconclusive about whether homebirth is "safer" and leads to better outcomes for mothers and babies. As Wilbur et al. (2015) have argued, the results of research on the outcomes associated with homebirth are highly dependent on both the methodology of the study as well as the geographic context of the research. Generalisable results are also difficult to achieve given the small absolute numbers of homebirth and the range of other variables that may affect birth outcomes (Chang and Macones 2011). In countries where homebirth is more common and integrated into well-developed healthcare systems (such as the U.K., the Netherlands and Iceland), research demonstrates similar or better outcomes for homebirth than hospital birth (Jonge et al. 2013). However, in the context of the United States, while some researchers argue that homebirths have better outcomes than hospital births (Johnson and Daviss 2005), others have found higher rates of neonatal and maternal risk associated with homebirth (Chang and Macones 2011). This latter research reflects and is reflected in the unequivocal support that the medical establishment, including the American College of Obstetrics and Gynecology and the American Medical Association, have historically given to hospital birth in the United States. These perspectives are also mirrored in policies regulating birth options across the United States, as licensure for direct entry midwives, who are the most likely to attend homebirths, is not available in the majority of states (MANA 2011). This leads to a situation in which "planned homebirth is not well supported in the United States by the government, professional organizations, the insurance industry, or society" (Boucher et al. 2009, 119). However, even within the context of the United States, while researchers publishing in medical journals often cite problems associated with homebirth, research in midwifery journals is strongly supportive of the practice based on birth outcomes (Johnson and Daviss 2005). This is largely due to the different ways that "safety" and "risk" are understood and measured across the different disciplinary perspectives (Scamell and Alaszewski 2012).

Across geographic contexts, one of the primary reasons that women choose to give birth at home, in most cases in spite of institutional and cultural pressure against it, is their disagreement with the biomedical model of birth and the associated interventionist approach that is taken in hospitals (Boucher et al. 2009; Holten and de Miranda 2016). This perspective reflects what has been termed a "midwifery" model of birth, which views birth as a holistic social, spiritual and embodied event, rejecting the mechanised body-as-object perspective of the biomedical model. From the perspective of the midwifery model, the woman is an active, agentic subject in the birth process, with the midwife accompanying her on her reproductive journey from the prenatal through the postnatal period (Malacrida 2015). As such, the notions of empowerment, knowledge and control are critical elements in discourses surrounding birth in this model (Worman-Ross and Mix 2013). For women who birth at home and the midwives who attend them, homebirth presents a "safer" option in that there is reduced likelihood of intervention, including caesarean section, induction, assisted vaginal delivery and the administration of pain medication or anesthesia (Jackson et al. 2012). In addition, women choosing to birth at home are often motivated by a desire to be able to exert greater levels of autonomy and control over the birth process; to have humanised, continuous care across the prenatal, birth and postnatal periods; and to be in a familiar environment in which they feel comfortable (Abel and Kearns 1991; Boucher et al. 2009; Jouhki 2012; Holten and de Miranda 2016). As Meredith and Hugill (2017, 10) summarise, "women are seeking places of emotional and physical safety for birth" and it is for this reason that the home is chosen.

While homebirth thus represents an alternative to the hegemonic understanding of women's health espoused in the biomedical model, it has not gone without critique from critical and feminist scholars. Rather, discourses of natural birth, like biomedical discourses, may be equally normative in some contexts, ensuring that women assume a neoliberal subjectivity as they materialise discourses of "personal choice", "responsibility" and "good motherhood" through risk management and the consumption processes involved in natural and/or homebirth (Rossiter 2017). Furthermore, these discourses may serve to naturalise the feminisation of both home and "nature" in ways that have historically supported confinement of women to these realms (Domosh and Seager 2001). In this way, our understandings of "natural" birth are also socially constructed and disciplining of the maternal subject (Mansfield 2008; Malacrida 2015). Uncomplicated and apolitical understandings of "home" in the homebirth discourse may be similarly problematic, as for many women home does not represent a place of safety, comfort, autonomy or control (Blunt and Dowling 2006; Longhurst 2008). In this way, as scholars have warned, homebirth is not a panacea to the problems associated with a biomedical approach to birth (Davis and Walker 2010).

Spatial dimensions of reproduction and birth

Within the field of geography, there is a growing interest in not only the place of birth, but in the socio-spatial dimensions of biological reproduction. This research shares the perspectives of medical anthropology and sociology in arguing for the centrality of reproduction to understanding social systems, norms and power relations. From this perspective, the social and spatial organisation of birth and other reproductive processes are not simply "natural" or "biological" but are instead constituted and structured by broader political, cultural and economic practices. As such, human reproduction can be positioned "at the very core of social theory" (Browner and Sargent 2011, 2) as it both "reflects and shapes core societal values and structures" (Sargent and Gulbas 2011, 290). Central to this perspective is the understanding that questions of health, including birth, are critical sites in the expression and employment of power and resistance (Cheyney 2008). As a result of this, according to Hardy and Wiedmer (2005, 9), "one of the most contested sites at which we can witness the clash of different ideological systems is the birthplace".

Within the last decade, feminist geographers have worked to develop an explicit understanding of both the central role of space and place in discourses of reproductive processes and the ways in which power is deployed through these spatialised processes. This literature draws on feminist geographers' conceptualisation of the body as a geographic space which reflects and embodies social structure and meaning. From this perspective, control over the reproductive body is critical to governance and the deployment of power at all scales (Davis and Walker 2010; Fannin 2013). Critical geographies of reproduction extend outside of the site of the body, however, as reproductive activities give meaning to places and spaces at a variety of scales. As such, reproducing bodies and processes are found to be in place (in the home, hospital, et cetera) and out of place (on the street, in workspaces) at different times and in different spaces through the life course (Longhurst 2008, 2009; Lane 2014). Furthermore, as Klimpel and Whitson (2016) have argued, reproductive processes also derive meaning from, and give meaning to, the broader spatialised discourses of modernity, development and nationhood.

While geographers have examined a number of topics related to reproduction, including pregnancy, lactation and mothering (Mahon-Daly and Andrews 2002; Madge and O'Connor 2006; Longhurst 2008, 2009; Boyer 2011; Fannin 2013), birth has been a key site of interest for geographers. As Abel and Kearns (1991, 832) argued over 25 years ago, birth is a critical site to view the expression of social power and, in particular, the ways that "patriarchy in the form of professional structures has extended control into the most intimate times and spaces of women's lives". Geographers have thus contributed to an explicit spatialisation of birth by considering the construction of various "landscapes of birth" (Fannin 2003, 514), which are alternately constituted as appropriate or inappropriate settings for this process, including hospital birthing and operating rooms, maternity clinics, the internet and the home (Abel and Kearns 1991; Fannin 2003; Longhurst 2008, 2009; Davis and Walker 2010; Emple and Hazen 2014; Klimpel and Whitson 2016). As Fannin (2003, 516) argues, such "landscapes of social reproduction are actively created, both discursively and materially through processes of transformation and cultural mobilization of spatial metaphors". As a result, birthplaces express and materially manifest struggles over institutionalised and alternative meanings, ideologies and understandings of bodies, identities and places.

Methodology

In this chapter, I draw on 13 in-depth interviews with women who planned homebirths. These interviews are part of data collected for a larger ongoing research project focusing more generally on the spatiality of homebirth. The interviews for this project took place in rural Appalachian Ohio in 2017. While Ohio as a whole has rates of homebirth just above the national average, like many other states homebirth and midwifery are not regulated in Ohio (MANA 2011; MacDorman et al. 2014). As such, homebirths in this state, while not illegal, take place outside of a framework of regulation and without legally recognised birth attendants. Moreover, in the rural region of Appalachian Ohio, mothers may encounter limited access to reproductive healthcare resources (for example there are no birthing centres in this region and local hospitals limit and/or do not allow vaginal birth after caesarean). At the same time, the Appalachian region more generally has a strong history of midwifery and homebirth, and as such represents a place in which homebirth has persisted historically even within a context of the increased medicalisation of childbirth (Buchanan et al. 2000).

Interviews were semi-structured and lasted between 90 and 150 minutes. Interviews took place in women's homes or public spaces, and women were often accompanied by their children, who at points spoke to me about their understandings of homebirth as well. Interviews were audio recorded and transcribed in their entirety. I used a modified grounded theory approach to analysis, with all interviews being analysed to develop initial or "open" codes rooted in the narratives of the participants. These open codes were used to construct broader, thematic coding categories, which were then applied to the transcripts through a process of focused coding (Charmaz 2006). The names of all participants referred to in this chapter have been changed to psuedonyms.

Eleven of the women interviewed for this research had experience giving birth both at home and in a hospital setting, and four of the women had given birth multiple times at home. Additionally, my sample included women who had planned homebirths and transferred to the hospital, as well as women who had planned and given birth at home. All of the research participants represented in this article identify as white, so in this way this chapter presents the experience of one particular demographic group of women, which itself reflects the geographic context of rural Appalachia in which it was conducted. However, in spite of the homogeneity of the research participants in terms of race and sexual orientation, the research participants were in extremely diverse economic situations. While some of the participants are professionals, a number of others rely on Medicaid for insurance and have extremely limited incomes. As such, the decision to have a homebirth presented a very different economic commitment for different women. As a researcher, I came to be interested in the topic of homebirth through my own experience giving birth both in a hospital and at home. As a result, my own experiences of homebirth provided a frame for my entry into this research and my subsequent analysis of my participants' experiences.

Resisting the biomedical model

In the remainder of this chapter, I focus on the ways that the place of home functions to enact and embody resistance in the narratives of women's birth experiences. During my interviews with mothers who planned homebirths, the biomedical, institutionalised model of birth came through very clearly as women's focus of resistance. This correlates with other research suggesting that many women engage in homebirth in an explicit rejection of institutional norms, medical authority and the biomedical model of birth (Boucher et al. 2009; Davis and Walker 2010; Holten and de Miranda 2016). The following comment from Sarah, a mother of six who planned two homebirths, clearly conveys this perspective:

> I could have done the VBAC [vaginal birth after caesarean] at the hospital. But again I know what comes with a hospital birth. And I feel like they put you on a clock. They want everyone to fit this box, and that's just not the way it works. And then I've also, through reading and different things, I have realised the circumstances you find yourself in can make your body go into labour or just back off of it. And I feel like when all these interventions and all these things, and the sterile room and people checking on you every few minutes, that's not the best way to have a healthy birth.

In this comment, Sarah establishes the hospital as a space of intervention, and rejects the biomedical notion that a "sterile room" is the best place to give birth. From her perspective, this institutional space disciplines bodies into a particular "box", a process which is at odds with her understanding of what produces a healthy birth.

For the majority of my interviewees, this perspective was developed through their previous experience with hospital birth, often supplemented with personal research on the history of birth. Katie connects her own hospital experience to the history of women's experience in hospitals to explain what she did not want in a birth experience:

We [women] are pushed to make decisions that weren't what we wanted or where birth was very traumatic on our bodies. Or the baby was separated right away, or people were pushed to have an epidural and to be induced or whatever. All those things were things, like, I didn't want. I wanted to maintain control over my body, especially after the experience of trying to labour naturally in a hospital. It's like a stressful environment and so many people. And they want to put an I.V. on you, and there's beeping monitors, and there's this thing around your belly and people are staring at you and it's bright. Like they're telling you what to do and what your body's feeling and stuff. That's so much not what I wanted.

(Katie, one planned homebirth with transfer to hospital)

Like Sarah, Katie richly describes the embodied experience of birth in the hospital to contextualise her desire for a homebirth. As Davis and Walker (2010, 388) argue, rather than being a neutral environment, the hospital becomes a place which "functions as a technology of biomedicine", challenging and marginalising the ability of women to materialise the birth process in an alternative framework.

Yet while many women seek an alternative to a biomedical birth model through choosing birth centres, midwife-assisted hospital birth and planned drug-free births, less than 1 per cent of women in the United States choose to birth at home (MacDorman et al. 2014). This decision very frequently involves not only resisting medical authority, but also incurring economic hardship and disregarding the advice and wisdom of friends and loved ones. As such, how does giving birth at home enable women to enact resistance in a unique way? How does the space of the home provide distinct opportunities to understand and embody an alternative birth experience? In the following sections, I will detail four ways that women engage the space of the home in their narratives of homebirth to express their resistance to biomedical norms: to normalise birth, to allow greater bodily autonomy and mobility, to enable greater control over space and to promote a reterritorialisation of birth.

Normalising birth

The first way in which home signals resistance to biomedical norms is through removing birth from the world of the exceptional or "out of the ordinary" that the hospital or birthing centre represent and bring it into the space of the everyday. In this way, women described the place of home as *normalising* the process of birth. As Sophia, a mother of two children who had both been born at home says, "I feel like in a lot of ways the Western view of birth is treated like it's an ailment or an illness that needs intervention. And the homebirth community treats it more of, you know, just like as a normal rite of passage". In the following comment, Tori, who has had three planned homebirths, argues that giving birth at home allowed for the process of birth to be just a part of her "normal" life:

I got to choose my private space in that house, which was our bedroom. And yeah, it was just like... it was just part of our daily life and not, I don't know. Like I don't really go to the hospital... I just wanted it to be like a really normal thing. Like a really normal part of my life, and that's, I feel like all my births were that way which is really nice.

From this perspective, the home is the space of the everyday, and by choosing to give birth at home, the event becomes a part of normal routine. Later in the interview, Tori goes on to describe that she wanted her children to be there as well so that they would understand that birth is a normal thing: "I want these guys to grow up knowing that it's normal", she says. The home was thus symbolic in constituting the birth as an everyday event.

Other research participants also narrate ways that they tried to keep these births as "normal" as possible, and describe the space of the home as facilitating this activity. As Amanda, a mother of four with three birthed at home, describes below, this meant going about other daily routines as much as possible.

Well, I needed to do something besides sit around and wait for the baby and cry that it wasn't here, because I was just emotionally exhausted, too. So, I made spaghetti and my mom was on the phone with my aunt, but she was watching the kids, too. And I didn't tell her that they [the contractions] were getting closer and closer because she was on the phone. And then they hit two minutes while I was still cooking dinner because it was spaghetti. You had to brown the meat and do everything.

For these mothers, the home permitted an opportunity to engage in everyday activities, thus further normalising birth. For other participants, normalising birth by bringing it into the space of the everyday was part of a broader process of normalising the presence of children in their lives, as Stephanie, a mother of three who had delivered one child at home, describes below:

Stephanie: All those things sort of fit together in like a normalising, you know, like sort of an attachment parenting philosophy, in normalising having children and children being part of, like an integrative part of life, and not just like family here, you know [Stephanie makes a gesture of a square with her hands to indicate a separate space]. ... its just like you can have your babies at home and you can bring your babies here and you can breastfeed your babies everywhere...
RW: So if we all do it our own way...
Stephanie: Our own way, everything becomes normal.

The home has the power to normalise, not simply because we spend time in it daily, but because it is a space of routine and habit (Wise 2000). It is clear from the above examples that the activity of birth itself does not have a single, inherent social meaning, but rather that the meaning of birth is contested by different communities. In these cases, choosing a different space for the birth helps to create a new meaning of it, as the space facilitates a redefinition of the activity as "normal". Furthermore, as hooks (1990) describes through her discussion of "homeplace", everyday activities in the home have the power to become political to the extent that they enable the creation of counter-hegemonic spaces and meanings. In this case, the normalcy of birth, children and family becomes a counterhegemonic reality.

Bodily autonomy and mobility

A second way in which my research participants described the home functioning as space of resistance in birth is by allowing them to have greater bodily autonomy, particularly through increased mobility. A desire for mobility in birth and the freedom to, as my participants expressed, "listen to your body" and "be free to follow the process", were part of every interview that I conducted. This contrasted with an expectation among women that in the hospital, even in the best of circumstances, bodily movement would be circumscribed and controlled.

In the homebirth experience, bodily autonomy began before the birth itself, as multiple women described not being "checked"—in other words, not having to submit to what they felt were unnecessary pelvic exams—during the prenatal period and birth. As Stephanie mentioned, "Nobody did any exams on me or checks on me and I was like, 'I think I'm ready to push', and they were like, 'go ahead and push'. You know it was very, very organic". During the birth itself, the ability to move freely and not be confined to a bed, to a single room or to the indoors were critical in descriptions of homebirth experiences. Christy, a mother of two with one planned homebirth, describes inhabiting "every inch" of her home during the labour process:

> I did a lot of up and down the stairs, in the shower, out of the shower, outside, inside. The story is that I laboured on every inch of the house and most of the three acres around it. I was just like wandering and moaning like a cow.

Similarly, Erin, a mother of three who had given birth to two at home, narrates how she was free to move about her house during labour even when this meant leaking amniotic fluid.

> I'm actually really like kind of hyper. Like I need to burn off some energy. So [I thought] maybe I'll walk or something. So of course, I decided to bake. I think I baked cookies. Yes, I baked cookies. Well the best part was I had this skirt on because I was like, "Oh, my water broke!" So I had to, like, go put this skirt on. And my brother and [husband] just followed me around the house because I was just leaking water everywhere.

As Kukla (2005) and Longhurst (2008) document, pregnant women's bodies are viewed as public spaces; that is, with pregnancy, women's bodies become subject to increased surveillance and public control. This is nowhere clearer than in the biomedical model of birth, wherein women's bodies come to be understood, at best, as simply a vessel for the unborn child or, at worst, as an obstacle from which the child needs to be freed (Davis-Floyd 2003; Kukla 2005). As such, within a biomedical model, women have historically been prohibited from basic activities such as walking, being upright and eating during the birth process, as these activities are understood as unnecessary for the removal of the foetus (by the obstetrician) (Davis-Floyd 2003). While these activities are increasingly allowed in hospital birth, they are nonetheless monitored, with the possibility of engaging in them at risk of revocation at any time. The women who participated in this research were keenly aware that the activities they engaged in while in labour—such as cooking, swimming in ponds, going grocery shopping and, very basically, moving with complete freedom from room to room—would not have been present as possibilities in a hospital context. For the women who participated in this research, however, control over their body was critical to their birth experience, and being at home enabled this control.

Furthermore, this ability to move freely and control their own bodies during the birth process was perceived as critical to the successful outcome of the birth itself. Participants described it as relaxing and allowing their bodies to do what they needed to do during the birth process. This is clear in Amanda's answer when asked what she enjoyed about being able to birth at home: "Being able to move around however I wanted. Because at the hospital as... soon as they started the Pitocin, they made me stay in bed from that point on. I like to be able to move around". Tori's experience is similar, as during each of her three homebirths, she found that she preferred to be alone and she was able to experience that:

> I feel like I really needed not to be touched and not be distracted and I got that. Like I could choose where I was... I didn't know I'd want to isolate myself, I had no idea. By the third, I was like, this is what I do. I find a cave and go into it, you know? I did that with all three of mine... I was able to follow my instincts as they arose and do what I needed to do. Which I think is, like, the best part.

For Tori and Amanda, the freedom to move as they felt they needed to was thus critical in what they saw as a healthy birth for both themselves and their babies.

Control over space

Similar to participants in other research on homebirth (Jouhki 2012; Worman-Ross and Mix 2013), the women who participated in this research saw the home not only as a space where they were able to control their own bodies and movements, but also a space which they were able to control more generally. This control took two forms: first, through mental and material place-making prior to the birth and, second, by controlling who was present in those spaces during the birth.

A number of the women with whom I spoke described the process of "creating spaces" for the birth ahead of time, as Katie does here:

> I think, like, we prepared, like, in terms of, like, spaces or I thought about it in terms of, like, spaces. Making sure that our bedroom was kind of clean and clear, that we had, like, a mattress protector on the bed... So, like, that idea, like, we could move space to space. We did the same thing in our living room. We rearranged the furniture and moved things out of the way... there was very much, like, the practical considerations of, like, creating the spaces.

This conscious process of "creating space" for the birth was important for many women, and involved moving furniture, preparing special lighting, having music ready and making arrangements for older children.

In addition to a physical process of making material preparations for the birth within the space of the home, women often prepared mentally for the spaces in which they would birth. The following excerpts from my interview fieldnotes document Erin's description of the mental process of preparing to give birth in a particular space:

> I asked her what she did to prepare her house for a homebirth and she said she spent a lot of time thinking about the space and imagining how the space would be for the birth. She described the house—she said it was a big farm house with a large, open, upstairs room with lots of light. She decided she would have the birth upstairs because of the space. She said that she also imagined that everyone else that she had invited to the birth would be downstairs, so that she could have the upstairs space. She repeated multiple times that "she thought a lot about the space". She told me that her midwife had her lay on the couch during her prenatal appointments and close her eyes and imagine herself giving birth in different places in the house, so she was able to really prepare herself for the space she wanted to give birth.

(Author's fieldnotes)

As Erin and Katie's experiences describe, giving birth at home allowed respondents to engage very explicitly in place-making focused around the birth experience. While mothers who give birth at home do so because they feel that home is an appropriate space for birth, they also continue to make and remake the materiality of home and its meaning in the process of preparing for and giving birth. As the home space is a critical site for the expression and creation of identity (Wise 2000; Blunt and Dowling 2006), reorganising and re-purposing these spaces for the birth process signal the ways in which both the home and identity are dynamically constituted around reproductive processes. As identities and social relations change during pregnancy and early motherhood, the material arrangement and use of the home reflects this. This is particularly visible in the event of a homebirth.

The research participants also appreciated being able to control not just the material objects in the birth space, but also the movements of others within that space. One component of this was being able to control who was in attendance at the birth, and at what points in the labour and birth process different people would be present. This is evident in Sophia's description of the many people who were at her house for her daughter's birth:

> And at this point, everyone was still, like, they were in a different room. They were just letting me have space. My husband was there and he was rubbing my back and pushing on my hips for me, cause that was really uncomfortable, and he was there, but they were just trying to stay out of the way. And then once I told them that my body was going and I couldn't stop it, then they all came in.

Many women wanted their children, mothers or other friends to be present at and a part of the birthing process, and were aware that many hospitals would have limitations in place in this regard. For other women, the ability to keep unknown people out of the space was equally critical. Rothman and Simonds (2005, 90) critique hospitals as places which women "do not control, and that are not controlled in their interests". As such, in spite of trends to make hospital rooms more comforting, welcoming and home-like (Fannin 2003), Rothman and Simonds (2005, 102) argue that "the choice to have a home-birth is not a choice of decorating schemes, familiar objects, the trivialities of a 'homey' atmosphere. It is a choice to control one's own body and space".

Re-spatialising birth

In her article on urban photographic exploration of abandoned maternity wards, Prescott (2009, 101) argues that birth is undergoing a "despatialization" as "it is becoming increasingly detached from physical, concrete spaces, from women's lived experience, and from practices of identity formation". Prescott traces the history of birth in the British context, describing its movement out of maternity units located within local community hospitals and into larger, centralised hospitals. She argues that this process not only "uproots" and "dislocates" birth from women's home communities, but that it also creates a spatialised experience that is not of the woman's choosing, but rather one that is imposed on her by the medical institution (Prescott 2009, 102). For Prescott (2009, 101), who is viewing birth spaces as spaces of memory, this despatialisation results in the silencing of birth experiences, as "narratives of birth are tied to places of birth". In contrast, as the previous section argues, the mothers interviewed for this research describe giving birth at home as a spatial tactic that allows for both material and subjective place-making processes to occur, providing a unique opportunity for women to enact resistance to this despatialisation of childbirth. These place-making activities reinforce and re-centre the importance of spatiality in both reproductive processes and in identity construction.

This intensely spatial nature of birth is visible in women's narratives in other ways as well. In addition to describing the home as a place which allowed them to intentionally control their birth experiences, almost all of my interview participants drew heavily on place to give meaning to their experiences. In the following excerpt, for example, Erin describes being closely connected to the place of the birth as a key element in her memory and experience of birth.

> I loved the fact that it was kind of my time. It was how my body, how I resonated. All those things. I wasn't on a clock, per se, like I felt like when I later had a hospital birth. I enjoyed the fact that it was my germs, my space, my stuff, my smells. All those things were very comforting. I knew where things were. I knew what steps to take and how I could move within the room, with my eyes closed. Those were all very comforting things for me.

Other research participants vividly and lovingly spatialised their birth narratives by remembering and describing not only the interior of the house itself, but also features of the natural environment within which they were located. When I asked Christy to describe her homebirth, for example, she began by locating it and herself in the place where she gave birth, stating, "It was a September evening, rural, all of the windows and doors were wide open. It was, like, humid, so there was that real sense of, like, the air being sort of thick, like the atmosphere was warm, moist and thick." Similarly, when asked about her strongest memory from her first homebirth, Sophia also locates the birth within the natural environment, as is evident from the following comment:

> They waited until it stopped pulsing to cut the cord; my husband cut the cord. We were in our living room. I put on my robe, I laid on my couch, I breastfed my baby for the first time. I looked up at the window, and I saw, it was one of the most powerful things I saw. I saw a praying mantis, a mama, who was swollen with eggs, sitting in that window. Got in the window while I was in labour, she stayed there for three more days, just like watching us, and then she jumped down into the bush, right by the window, and laid her eggs. It was so cool... So for days, I kept looking up different words for praying mantis, or in Latin, and all the names were terrible. So [my daughter's middle name] became Rain. It was raining that day.

These narratives, as well as those cited above describing making cookies, frying ground beef, walking up and down stairs, finding a "cave" to birth in and moving freely through the house and yard, all reflect the intensely spatialised experience of birthing at home.

While the sterility and uniformity of the hospital space is one of the characteristics that marks it as a "safe" space to many, this uniformity can also create a disconnection from place (Prescott 2009). In contrast, being at home for the birth allowed the mothers interviewed for this project to experience and memorialise a meaningful connection to birthplace. Sophia expresses this perspective eloquently:

> Home is where your heart is. And my heart has grown tremendously having children. And I love that we add that extra layer of our home experience by having them been born there. Like my daughter will proudly point to the part on the living floor and tell people, this is where I was born. And this is where my brother was born.

Rothman and Simonds (2005, 97) argue that in hospitals "women's bodies and their birth experiences becomes a matter of territory" and that "the environment is itself a form of colonization". For many women, including those interviewed here, birthing at home moves women out of this environment of colonisation and into spaces of their own creation.

Conclusion

In this chapter I have argued that giving birth at home allows women to resist the norms associated with medicalised birth and institutional control in unique ways, and that the space of the home is central to this. I describe four ways that the space of the home plays a role in this process: by normalising birth, supporting women's mobility and bodily autonomy during birth, permitting women increased control over space during birth and re-spatialising the experience of birth. According to Routledge (1997, 71) spaces of resistance may be created to the extent that actors attempt to "establish (however temporarily) social spaces and socio-spatial networks that are insulated from control and surveillance". Similarly, Hamdan-Saliba and Fenster (2012, 203) describe resistance as the employment of "pro-active tactics of power", which function to "create alternative spaces; enable the manipulation of social and cultural codes; and create the home as a space of independence". In the material discussed in this chapter, birthing at home can thus be seen to function as a pro-active tactic in that it enables women to establish spaces that support a non-normative meaning and experience of birth.

The findings presented in this chapter both draw on and support feminist and geographic scholarship on space and resistance which highlights the importance of both the home and the body as critical sites of envisioning and enacting alternative politics (hooks 1990; Blunt and Dowling 2006). These findings also extend scholarship on homebirth that focuses on reasons why women choose to birth at home, or the benefits or problems associated with such births. While some of the material presented above is framed by the women themselves as reasons or benefits, the focus of this chapter is not evaluating the process of homebirth itself, but better understanding how it functions discursively and materially to present an alternative to normalised meanings and experiences of birth, and how the place of the home is a critical component of this function. Certainly, as previous research indicates, not all women have positive experiences with homebirth, nor does home function for all women as a space where alternative meanings supporting control and independence can be established (Blunt and Dowling 2006; Longhurst 2008; Davis and Walker 2010). However, for the women interviewed here, the home allowed women to envision, articulate and experience a space of refusal to medicalised and institutionalised norms.

References

Abel, Sally, and Robin A. Kearns. 1991. "Birth Places: A Geographical Perspective on Planned Homebirth in New Zealand." *Social Science & Medicine* 33(7): 825–834.

Betrán, Ana Pilar, Jianfeng Ye, Anne-Beth Moller, Jun Zhang, A. Metin Gülmezoglu, and Maria Regina Torloni. 2016. "The Increasing Trend in Caesarean Section Rates: Global, Regional and National Estimates: 1990–2014." *PLoS ONE* 11(2): 1–12.

Blunt, Alison, and Robyn Dowling. 2006. *Home*. New York: Routledge.

Boucher, Debora, Catherine Bennett, Barbara McFarlin, and Rixa Freeze. 2009. "Staying Home to Give Birth: Why Women in the United States Choose Homebirth." *Journal of Midwifery & Women's Health* 54(2):119–126.

Boyer, Kate. 2011. "'The Way to Break the Taboo Is to Do the Taboo Thing' Breastfeeding in Public and Citizen-Activism in the UK." *Health & Place* 17(2): 430–437.

Browner, Carole H., and Carolyn F. Sargent. 2011. "Toward Global Anthropological Studies of Reproduction: Concepts, Methods, Theoretical Approaches." In *Production, Globalization, and the State: New Theoretical and Ethnographic Perspectives*, edited by Carole H. Browner and Carolyn F. Sargent, 1–18. Durham, NC: Duke University Press.

Buchanan, Patricia, Vicky K. Parker, and Ruth Hopkins Zajdel. 2000. "Birthin' Babies: The History of Midwifery in Appalachia." Paper presented at the Annual Conference of the Women of Appalachia: Their Heritage and Accomplishments. Zanesville, OH.

Chang, Jen Jen, and George A. Macones. 2011. "Birth Outcomes of Planned Homebirths in Missouri: A Population-Based Study." *American Journal of Perinatology* 28(07): 529–536.

Charmaz, Kathy. 2006. *Constructing Grounded Theory: A Practical Guide Through Qualitative Analysis*. London: Sage.

Cheyney, Melissa J. 2008. "Homebirth as Systems-Challenging Praxis: Knowledge, Power, and Intimacy in the Birthplace." *Qualitative Health Research* 18(2): 254–267. https://doi.org/10.1177/1049732307312393.

Davis, Deborah, and Kim Walker. 2010. "The Corporeal, the Social and Space/Place: Exploring Intersections from a Midwifery Perspective in New Zealand." *Gender, Place & Culture* 17(3): 377–391.

Davis-Floyd, Robbie. 2003. *Birth as an American Rite of Passage*. Berkeley: University of California Press.

Domosh, Mona, and Joni Seager. 2001. *Putting Women in Place: Feminist Geographers Make Sense of the World*. New York: Guilford Press.

Emple, Hannah, and Helen Hazen. 2014. "Navigating Risk in Minnesota's Birth Landscape: Care Providers' Perspectives." *ACME: An International Journal for Critical Geographies* 13(2): 352–371.

Fannin, Maria. 2003. "Domesticating Birth in the Hospital: 'Family-Centered' Birth and the Emergence of 'Homelike' Birthing Rooms." *Antipode* 35(3): 513–535.

Fannin, Maria. 2013. "The Burden of Choosing Wisely: Biopolitics at the Beginning of Life." *Gender, Place & Culture* 20(3): 273–289.

Hamdan-Saliba, Hanaa, and Tovi Fenster. 2012. "Tactics and Strategies of Power: The Construction of Spaces of Belonging for Palestinian Women in Jaffa–Tel Aviv." *Women's Studies International Forum* 35(4): 203–213.

Hardy, Sarah, and Caroline Wiedmer. 2005. "Introduction: Spaces of Motherhood." In *Motherhood and Space: Configurations of the Maternal through Politics, Home, and the Body*, edited by Sarah Hardy and Caroline Wiedmer, 1–14. New York: Palgrave Macmillan.

Holten, Lianne, and Esteriek de Miranda. 2016. "Women's Motivations for Having Unassisted Childbirth or High-Risk Homebirth: An Exploration of the Literature on 'Birthing Outside the System.'" *Midwifery* 38 (July): 55–62.

hooks, bell. 1990. *Yearning: Race, Gender, and Cultural Politics*. Boston MA: South End Press.

Jackson, Melanie, Hannah Dahlen, and Virginia Schmied. 2012. "Birthing Outside the System: Perceptions of Risk amongst Australian Women Who Have Freebirths and High Risk Homebirths." *Midwifery* 28(5): 561–567.

Johnson, Kenneth C., and Betty-Anne Daviss. 2005. "Outcomes of Planned Home-births with Certified Professional Midwives: Large Prospective Study in North America." *BMJ* 330(7505): 1416.

Jonge, Ank de, Jeanette A. J. M. Mesman, Judith Manniën, Joost J. Zwart, Jeroen van Dillen, and Jos van Roosmalen. 2013. "Severe Adverse Maternal Outcomes among Low Risk Women with Planned Home versus Hospital Births in the Netherlands: Nationwide Cohort Study." *BMJ* 346 (June): f3263.

Jouhki, Maija-Riitta. 2012. "Choosing Homebirth–The Women's Perspective." *Women and Birth* 25 (December): e56–61.

Klimpel, Jill, and Risa Whitson. 2016. "Birthing Modernity: Spatial Discourses of Cesarean Birth in São Paulo, Brazil." *Gender, Place & Culture: A Journal of Feminist Geography* 23(8):1207–1220.

Kukla, Rebecca. 2005. *Mass Hysteria: Medicine, Culture, and Mothers' Bodies*. Lanham, MD: Rowman & Littlefield Publishers.

Lane, Rebecca. 2014. "Healthy Discretion? Breastfeeding and the Mutual Maintenance of Motherhood and Public Space." *Gender, Place & Culture* 21(2): 195–210.

Longhurst, Robyn. 2008. *Maternities: Gender, Bodies and Space*. New York: Routledge.

Longhurst, Robyn. 2009. "YouTube: A New Space for Birth?" *Feminist Review* 93(1): 46–63.

MacDorman, Marian F., T. J. Mathews, and Eugene Declercq. 2014. "Trends of Out-of-Hospital Births in the United States, 1990–2012." *NCHS Data Brief 144.* Washington, DC: US Department of Health and Human Services.

Madge, Clare, and Henrietta O'Connor. 2006. "Parenting Gone Wired: Empowerment of New Mothers on the Internet?" *Social & Cultural Geography* 7(2):199–220.

Mahon-Daly, Patricia, and Gavin J. Andrews. 2002. "Liminality and Breastfeeding: Women Negotiating Space and Two Bodies." *Health & Place* 8(2):61–76.

Malacrida, Claudia. 2015. "Always, Already-Medicalized: Women's Prenatal Knowledge and Choice in Two Canadian Contexts." *Current Sociology* 63(5): 636–651.

MANA (Midwives Alliance of North America). 2011. "Direct Entry Midwifery State-by-State Legal Status." Midwives Alliance of North America. https://mana.org/pdfs/Statechart-05-11-11.pdf.

Mansfield, Becky. 2008. "The Social Nature of Natural Childbirth." *Social Science & Medicine* 66(5):1084–1094.

Meredith, Dawn, and Kevin Hugill. 2017. "Motivations and Influences Acting on Women Choosing a Homebirth: Seeking a 'cwtch' Birth Setting." *British Journal of Midwifery* 25(1): 10–16.

Prescott, Holly. 2009. "Birth-Place." *Feminist Review*, 93: 101–108.

Rossiter, Kate. 2017. "Pushing Ecstasy: Neoliberalism, Childbirth, and the Making of Mama Economicus." *Women's Studies* 46(1): 41–59.

Rothman, Barbara Katz, and Wendy Simonds. 2005. "The Birthplace." In *Motherhood and Space: Configurations of the Maternal through Politics, Home, and the Body,* edited by Sarah Hardy and Caroline Wiedmer, 87–104. New York: Palgrave Macmillan.

Routlege, Paul. 1997. "A Spatiality of Resistances: Theory and Practice in Nepal's Revolution of 1990." In *Geographies of Resistance*, edited by Steve Pile and Michael Keith, 68–86. New York: Routledge.

Sargent, Carolyn, and Lauren Gulbas. 2011. "Situating Birth in the Anthropology of Reproduction." In *A Companion to Medical Anthropology*, edited by Merrill Singer and Pamela I. Erickson, 289–303. Malden, MA: Blackwell.

Scamell, Mandie, and Andy Alaszewski. 2012. "Fateful Moments and the Categorisation of Risk: Midwifery Practice and the Ever-Narrowing Window of Normality during Childbirth." *Health, Risk & Society* 14(2): 207–221.

Wilbur, M. B., S. Little, and L. M. Szymanski. 2015. "Is Homebirth Safe?" *New England Journal of Medicine* 373(27): 2683–2685.

Wise, J. Macgregor. 2000. "Home: Territory and Identity." *Cultural Studies* 14(2): 295–310.

Worman-Ross, Kathryn, and Tamara L. Mix. 2013. "'I Wanted Empowerment, Healing, and Respect': Homebirth as Challenge to Medical Hegemony." *Sociological Spectrum* 33(5): 453–481.

Part III
Politics

9 Birth and biopolitics

Maternity migration, birthright citizenship and domopolitics in Hong Kong

Robert Kaiser

January 2014. During a shuttle van ride from Shenzhen to the Hong Kong airport, we cleared the Chinese border control station and entered the Hong Kong checkpoint. A female border guard approached the van and ordered a middle-aged woman to remove her purse from her lap and open her coat. As of 1 January 2013, a ban was placed on mainland maternity migration, and the border guard was checking for signs of pregnancy. The woman protested that she was not from China, but from Taiwan, and was returning home. The border guard insisted that she comply and, following a brief shouting match, the woman did. Her body shape left the border guard uncertain, and the woman and her luggage were removed from the van. These practices have become routine—Chinese-looking women who appear to be of childbearing years regularly removed from cross-border vans and their bodies and baggage searched for signs of pregnancy. This continues today.

Introduction

Hong Kong became a Special Administrative Region (SAR) of China on 1 July 1997 under the "one country, two systems" policy, which provided Hong Kong a degree of political, legal and territorial autonomy for a period of 50 years. This meant that Hong Kong was not forced to adopt China's laws and policies, including its "one child policy". The imposition of this biopolitical mechanism of reproductive control would have made little difference for Hong Kong's birthrate, since at the time of the "handover" from the U.K. to China Hong Kong had one of the lowest birth rates in the world. Indeed, its rapidly ageing population and shrinking labour force were cited as the most serious demographic threats facing the territory. In the eyes of its population and economic planners, Hong Kong needed more, not fewer, babies (Task Force on Population Policy 2003).

In Shenzhen and other border regions near Hong Kong, an increasing number of mainland Chinese couples took advantage of the opportunity to give birth to a second child in Hong Kong and avoid the restrictions and penalties associated with China's one child policy. Beyond the desire for a second child, children born in Hong Kong would automatically receive the "right of abode",[1] essentially birthright citizenship, and many parents believed this would provide their child the opportunity for a brighter future. Having the right of abode entitled the child, though not the mainland parents, to education, healthcare, residency and easier access to housing and job opportunities in Hong Kong and potentially abroad. There was also a perception that healthcare services in Hong Kong were better for expectant mothers.

Given the rapid greying of Hong Kong's population, maternity migration from the mainland should have been welcomed. At the beginning of the 2000s, Hong Kong's population, health and education professionals did view mainland maternity migration relatively positively, something that would benefit Hong Kong if managed properly. From the outset, however, more nativistic public and political voices were raised in opposition to this practice, presenting it as a threat to Hong Kong and its people. As the numbers of mainland births increased, these nativistic voices grew louder and more extreme, and before the decade was out even Hong Kong's demographic administrators had changed their stance. By 1 January 2013, an outright ban was imposed on the practice.[2]

From the perspective of governmentality and biopolitics, mainland maternity migration should have been welcomed and promoted as a solution to Hong Kong's rapidly shrinking younger generation and workforce and rapidly growing elderly population. Why did it have the opposite effect? And what might this tell us about homeland nationalism and birthright citizenship more generally? In this chapter, I use Didier Bigo's work on the governmentality of unease and William Walters' research on domopolitics, as well as my own study of nationalism, territoriality and homeland politics to explore the issues of maternity migration, birthright citizenship and the nativism and exclusionary nationalistic practices they often trigger. I then focus on the relationship among birthright citizenship, maternity migration and the "fetal citizen" (Wang 2017) and the production of "accidental" or "alien" citizens (Nyers 2006; Cisneros 2013). In the third section of the chapter, I provide a detailed critical analysis of mainland maternity migration to Hong Kong.

Governmentality of unease, immigration, domopolitics

Foucault (2007, 18) opens his lectures on governmentality with a discussion of space as an apparatus of security and argues that opening up towns to greater circulation with the outside world was the principal objective of government: "it was a matter of organizing circulation, eliminating its dangerous elements, making a division between good and bad circulation, and maximizing the good circulation by diminishing the bad". In this text, "good and bad circulation" are dealt with unproblematically: better ventilation to improve hygiene and minimise disease and poor health; greater external trade and increased wealth accumulation while minimising crime and theft. What Foucault fails to consider here is how the dividing practices bordering "good" and "bad" circulation are performed, who is empowered to perform them and what the consequences are for those included and excluded by such borderings.[3]

Governmentality of unease

These questions are central to security discourse surrounding the processes of immigration and the circulation of human bodies across political borders. In an important extension of Foucault, Bigo's (2002) work on "the governmentality of unease" is particularly useful as a starting point for answering these questions. Bigo asks why immigrants are continually framed as problems of security, and rejects the most frequent answer given (i.e., because they *are* a threat to security), by pointing out the myriad studies that conclusively demonstrate immigrants are positive socially, economically and culturally, and are less of a security risk than the "home" population. In other words, by most of the objective criteria identified by Foucault, immigration and immigrants should be considered "good" circulation. Bigo argues, rather, that immigration is treated as a problem of security, "when it is presented as such by the professionals in charge of the management of risk and fear...(who) transfer the legitimacy they gain from struggles against terrorists, criminals, spies and counterfeiters toward other targets, most notably transnational political activists, people crossing borders, or *people born in the country but with foreign parents*" (Bigo 2002, 63, emphasis added). More fundamentally, the treatment of immigration as a threat "is based on our conception of the state as a body or a container for the polity. It is anchored in the fears of politicians about losing their symbolic control over the territorial boundaries" (Bigo 2002, 65). The imaginative geography that essentialises the nation-state as a political body and homeland also naturalises opposition to "foreigners" and immigrants as a "threat to the homogeneity of the people" (Bigo 2002, 67).

One critique that may be lodged against Bigo's depiction of the governmentality of unease is the role played by professional managers of unease. He argues that it is the professionals who are in a privileged position (have the legitimacy) to present immigrants as threats, or not, and that amateurs or the public at large struggle to have their views heard and adopted. While there are undoubtedly circumstances in which this is the case, the role of biopolitical experts seems less central in the context of anti-immigrant nativism. Where national identity materialises in and through a set of everyday citational practices that naturalise both the nation as a biological community centred around a belief in common ancestry and the state as a homeland perceived as the birthplace of the "organic" nation, members of this homeland nation assume for themselves the position of experts with the cultural and political capital to speak in the name of national community and homeland state. They frequently lead the charge against what they perceive as threats to the "organic" nation's control over its homeland. This is particularly likely to be the case among those self-proclaimed members of the homeland nation who have lost status or seen the future prospects for themselves and their children in the national homeland decline as a result of "foreign competition" (Somers 2008, 132–143). It is these people who often organise against immigrants as those who are cheating the system, stealing what rightfully belongs to members of the homeland nation, who don't know their place in someone else's homeland. In the case of mainland maternity migration to Hong Kong, as I detail below, it was popular unease that mobilised politicians to restrict and finally ban the practice, while the professional managers of biopolitical unease in city government treated maternity migration as something that could be managed and regulated for the benefit of Hong Kong, until they were forced by growing public hostility to take a more exclusionary stance.

Domopolitics

William Walters (2004, 241) adds a homeland politics or "domopolitics" dimension to the dividing practices associated with biopolitics in his study of Britain's 2002 White Paper "Secure Borders, Safe Havens":

> Domopolitics implies a reconfiguring of the relations between citizenship, state, and territory. At its heart is a fateful conjunction of home, land and security. It rationalizes a series of security measures in the name of a particular conception of home...[that] has powerful affinities with family, intimacy, place: the home as hearth, a refuge or a sanctuary in a heartless world; the home as *our* place, where we belong naturally, and where, by definition, others do not; international order as a space of homes—every people should have (at least) one; home as a place we must protect. We may invite guests into our home, but they come at our invitation; they don't stay indefinitely. Others are, by definition, uninvited. Illegal migrants and bogus refugees should be returned to 'their homes'... Hence domopolitics embodies a tactic which juxtaposes the 'warm words' (Connolly, 1995, p. 142) of community, trust, and citizenship, with the danger words of a chaotic outside—illegals, traffickers, terrorists; a game which configures things as 'Us vs. Them'.

Walters argues that domopolitics is not simply a return to the homeland politics and exclusionary nationalism of the late nineteenth to mid-twentieth century, since a globally integrated and interdependent economic system has replaced the discrete national economies of the earlier era. The government's task is to manage interstate circulation, "to attract and channel flows of resources, whether investment, goods, services, and now flows of (the right kind of) people into one's territory" (ibid., 244).

Walters makes the case that domopolitics—a new form of biopolitics in which the state is narrated and enacted as a home(land) to be protected and defended—emerged after 9/11 and exists in tension with the more open neoliberal governmentality that manages the state as a business, and that this tension between these forms of biopolitics is especially apparent in the context of immigration. While countries continue to welcome and recruit immigrants who are viewed as beneficial to the state (i.e., "the right kind of people"), they increasingly restrict and curtail uninvited immigrants—the undocumented, refugees, etc.—who are imagineered (i.e., imaginatively engineered) as "the wrong kind of people"—criminals or potential terrorists, culturally unassimilable and/or economic drains on state and society. Nationalists conjure these uninvited and unwelcome guests in the home (land) as a visceral threat to the national body politic. Walters also identifies a more assertive promotion of loyalty to the state as a home, which represents the second meaning of domo: to tame, conquer, domesticate. Together, these approaches to governmentality,

> produce a particular politics of mobility whose dream is not to arrest mobility but to tame it; not to build walls, but systems capable of utilizing mobilities, tapping their energies and in certain cases deploying them against the sedentary and ossified elements within society; not a generalized immobilization, but a strategic application of immobility to specific cases coupled with the production of (certain kinds of) mobility.
>
> (ibid. 2004, 248)

We might question this last aspect of domopolitics, given the recent penchant for wall-building in the U.S., the E.U., Israel and elsewhere. However, according to Wendy Brown (2010, 24), wall-building is a reaction to the fact that we have entered a "post-Westphalian" world: "it is the weakening of state sovereignty, and more precisely, the detachment of sovereignty from the nation-state, that is generating much of the frenzy of nation-state wall building today. Rather than resurgent expressions of nation-state sovereignty, the new walls are icons of its erosion".

Most of the literature that has utilised domopolitics to date has focused on asylum seekers (Ingram 2008; Darling 2011, 2014; Hynek 2012; though see Titley 2012), and the increasing tendency to label them as "bogus refugees" or "economic migrants"—as "takers" rather than "givers". The domopolitical emphasis on asylum seekers as one of the most vulnerable immigrant groups is certainly warranted, and brings into sharp focus the dividing practices separating "good" migrants (economically beneficial labour and "genuine" refugees needing protection) from "bad" migrants (undocumented workers, "bogus" refugees, terrorists). However, the focus on asylum seekers has not allowed for a more comprehensive interrogation of the arts of governing the state as a homeland. Immigration more generally, as Bigo has argued, is a ubiquitous target of professional "managers of unease". As I detail in the following section, maternity migrants are even more likely to trigger exclusionary homeland nationalism, particularly in states with birthright citizenship.

Additionally, much of the domopolitical literature to date defines the homeland population on the basis of citizenship. While this will frequently be the basis on which a domopolitical dividing line is drawn in official documents, and although at the most fundamental level citizenship does provide that minimum protection which Arendt described as "the right to have rights" (Arendt 1951; Somers 2008), the nationalist biopolitical discourse that performatively materializes the nation as an organic entity rooted in the soil of its homeland tends to deploy much more exclusionary bordering practices that produce a multiplicity of threatening Others within, frequently imagining them as all the more threatening *because* they have citizenship. The nation and homeland are typically portrayed within this discourse as far more ancient than the citizenry and state, so that domopolitics is based on the fear and anxiety of the self-defined homeland nation, whose members feel that their preeminent position in the homeland is slipping away, that the government is not serving their interests first and foremost and that their children's future is less bright and more uncertain (Bigo 2002, 65; Somers 2008). Domopolitics and the managers of unease tap into these homeland nationalist fears and anxieties by targeting not only non-citizen immigrants, but also citizens exteriorised from the homeland nation as racial, ethnic and sexual minorities, as well as "alien" and "accidental" citizens (Nyers 2006; Ngai 2007) and birthright citizenship more generally. The ways in which homeland nationalists target maternity migration and birthright citizenship highlights how reproduction is central to contemporary biopolitics, something that the domopolitics literature to date—with its emphasis on the citizen/non-citizen divide—has largely overlooked. Growing homeland nationalist unease over maternity migration, and the recent attacks on and rollback of birthright citizenship, are the subjects of the following section.

Maternity migration and birthright citizenship as threats to the homeland nation

Babies born to immigrant parents (documented and undocumented) who have not yet received citizenship or permanent resident status are the frequent targets of homeland nationalists, nativists and xenophobes. Maternity migration, even more than refugee migration itself, is likely to trigger anti-immigrant nativism and a rise in domopolitics, especially but not exclusively in states with birthright citizenship.[4] The imagery of maternity migrants producing "anchor babies" has been centrally featured by exclusionary nationalists seeking to align citizenship with more restrictive understandings of the homeland nation. In the United States, maternity migration and birthright citizenship sparked heated debates during the 2005 efforts to repeal the 14th Amendment (i.e., The Citizenship Reform Act), and were featured in the 2016 presidential contest, in which Donald Trump declared that "anchor babies" born to "illegal immigrants" were not entitled to citizenship under the 14th Amendment (Farley 2015). Birthright citizenship for babies born to couples who are not citizens or permanent residents has been rescinded in Ireland (2005), New Zealand (2005), Australia (1986), and the U.K. (1981). According to Ngai (2007, 2530), "in each case, the changes were made at least partly, if not primarily, in response to popular nativist sentiment against non-white immigrants".

In Ireland, for example, which was the last E.U. member country with unrestricted *jus soli* citizenship, the 2004 referendum "passed with a nearly 80 percent majority," and while the debate focused on the "intensity of immigrants' connections to the Irish State...the government clearly employed racist and xenophobic tropes to link the presence of the immigrants to a pressing contemporary political issue: the crisis in the health-care system" (Mancini and Finlay 2008, 575, 583). In particular, the "costs of births to 'illegal alien mothers' and the threat of disease", as well as a charge which was later discredited that foreign pregnant women were overwhelming Dublin's maternity hospitals, were used to mobilise public opinion against unrestricted birthright citizenship (ibid., 579–583). Childbearing among migrant women was presented by the government as threatening "the possibility of creating a desirable future for Ireland and 'legitimate' Irish people" by the demands they and their children would make on social welfare, healthcare, education and employment (Luibhéid 2013, 161). As I detail below, these claims are eerily similar to those deployed against mainland maternity migration in Hong Kong.

In places where citizenship is still automatically conferred at birth regardless of the status of the parents, mother and child are frequently treated as threats to the national body politic, and subject to derision and abjection. In the U.S., for example, birthright citizenship is continually presented by nativists as "the source of...the illegal alien invasion" and "national political discourses surrounding 'anchor babies' and 'alien maternity' constitute the always-already racialized 'alien' subject as dangerous and perverse" (Cisneros 2013, 290, 292). While the unborn of the homeland nation are imagined to comprise the "true" Americans and are treated as "fetal supercitizens", the foetuses of non-status pregnant women "are constituted as racialized, anticitizen 'anchor babies'" (ibid., 297; see also Wang 2017). The dividing practices that produce the foetus as "alien" or "accidental" citizen (Berlant 1994; Nyers 2006) and migrant mother and child as an "invasion force" that creates the conditions of possibility for imagining pregnant women and the foetuses they carry as an existential threat to the "organic" nation in its homeland.

Nyers (2006, 24) uses Paul Virilio's insight that the accident is immanent to any technological object (e.g., derailment as immanent to the creation of the railroad) to explore the discourse of "accidental citizenship"—referencing the babies born to non-status parents in countries with unrestricted birthright citizenship as "the abject counterpart to the essential citizen". Here, birthright citizenship is the biopolitical object to which "accidental citizenship" is immanent. Looking at the use of accidental citizenship as justification for the internment of Japanese-Americans during WWII, and for the banishment of

Yaser Esam Hamdi (born in the U.S. to Saudi citizens and who was later turned in as a Taliban fighter in Afghanistan and treated as an enemy combatant), Nyers (2006, 26–32) argues that,

> If birthright is always at risk of being exposed as a birth accident, then the sovereign body politic equally risks being revealed as arbitrary and capricious...What is occurring is no less than a complex process of unmaking citizenship, of making something natural into something foreign. Foreignness is a symbolic marker that the nation attaches to the people we want to disavow, deport, or detain because we experience them as a threat. What is at stake is not the alien/citizen divide...but the capacity to decide upon and enforce the exception; that is, to expose something that is considered to be natural and normal (e.g., birthright citizenship) as arbitrary, accidental, and foreign.

Although few states allow for unrestricted birthright citizenship, with the exception of naturalisation procedures, all citizenship is fundamentally birthright citizenship. Since there is nothing natural about *jus sanguinis* or blood ties to a nation or state, citizenship is always tied to place of birth—whether it is second or third generation in place, or birth to those who were born in a place that is imagineered as their ancestral homeland. Given this, the ability of managers of biopolitical unease to revoke this birthright for some—to recast it as an accident of birth—or even to engage in the practice of "backward uncitizening" (Cisneros 2013) makes all citizenship less permanent, more tenuous and makes all citizens more vulnerable. Still, it is important to remind ourselves of the unevenness of power and privilege here: those whose claims to native status through the dividing practices that performatively naturalise the connection between blood ties and "essential" citizenship are able to stabilise their own hyphenated connection (nation-state) by undermining or destabilising the status of those cast in the role of the alien, the immigrant, the accidental citizen, the Other.

In the discussion below, I trace maternity migration from mainland China to Hong Kong, the issue of birthright citizenship and the ways in which babies born to two mainland Chinese parents—abjectly referenced as the "double-nons"—triggered unease and the rise of domopolitics among nativistic Hong Kongers, and were quickly rebordered from "good migrants" helping to preserve the future of Hong Kong to "bad migrants" representing an existential threat to Hong Kong and its "native" people.

Maternity migration and domopolitics in Hong Kong

As noted in the introduction, Hong Kong's special status under the "one country, two systems" policy that returned it to China meant that Hong Kong would not be required to adopt China's laws, including China's "one child policy", for a period of 50 years.[5] Adopting the "one child policy" in

Hong Kong would have made little sense in any event, since the territory had among the lowest birthrates in the world at the time, with a Total Fertility Rate (TFR) of 0.93 in 2001 (Task Force on Population Policy 2003, 2). The most critical demographic concerns facing Hong Kong's city and regional planners were the rapidly ageing population and shrinking workforce. In addition to fears about the economic costs of a large and growing elderly population, more immediate problems such as the underutilisation of maternity wards and the under-enrolment of school children in K-12 classrooms threatened the closure of facilities and the laying off of staff.

On the other side of the border in mainland China, rapidly rising incomes and the continuing desire among many couples to avoid China's one child restrictions made maternity migration to Hong Kong attractive. Additionally, many mainland couples desired to give birth in Hong Kong, since the baby would receive the "right of abode"—essentially birthright citizenship, which would provide the child with access to residence, healthcare and education in Hong Kong. Many also believed that maternal healthcare in Hong Kong was better than in mainland China.

Given the demographic challenges facing Hong Kong, mainland maternity migration and the increasing childbirth in Hong Kong that resulted should have been viewed as "good circulation". Additionally, mainland pregnant women paid higher rates for maternity care than Hong Kong residents, and many of the babies with birthright citizenship would be raised and educated in mainland China, but would come to study and work in Hong Kong as they grew older, helping to slow the greying of the population and the shrinking workforce. Mainland couples wishing to give birth to a second child in Hong Kong also tended to be highly educated, mostly working as managers and professionals, and with high incomes (Census and Statistics Department 2011). According to Walters' domopolitical framework, these were exactly the "right kinds of people" that Hong Kong should have wanted to attract. Yet just the opposite happened. Why were the babies born in Hong Kong to mainland couples exteriorised as the "wrong kind of people", as an existential threat that must be stopped at all costs?

The early 2000s: Maternity migration as "good circulation"

At the beginning of the 2000s, few concerns were raised by Hong Kong's population policy experts about mainland maternity migration. The 2003 *Report of the Task Force on Population Policy*, which was commissioned to identify population challenges and recommend policies to address them, makes no mention of births to mainland pregnant women (Task Force on Population Policy 2003). The focus of this report, as noted above, was on the extremely low birth rates, the rapidly ageing population, the shrinking labour force, and the threats these represented to Hong Kong's continued social wellbeing and economic competitiveness. The recommendations made would all suggest that maternity migration from the mainland would be seen as "good circulation" that would help to address these demographic problems. For example, the report recommended that Hong Kong do everything possible to maximise the number of children from mainland China coming to Hong Kong on the One-Way Permit (OWP) Scheme, a system put in place for family reunification of Hong Kong residents with their children and spouses living in China. The report argued that these children should relocate to Hong Kong at the youngest possible age, in order to facilitate their social integration as well as to prepare them to succeed in Hong Kong's higher education system and its knowledge-based economy (ibid., 48). The report also recommended pronatalist policies that could stimulate the birthrate in Hong Kong (ibid., 60–62).

Nonetheless, beginning in 2003, and driven by growing public concerns, members of Hong Kong's Legislative Council (LegCo) began interrogating Hong Kong's professional managers of biopolitical risk about mainland pregnant women giving birth in Hong Kong. The main concerns raised by LegCo members were the limitations this might put on public healthcare access for locals and the rate of defaults on debts owed. The responses by administrative officials were mainly reassuring. For example, at a hearing held on 26 November 2003 the Secretary of Security responded that,

> The number of childbirths by mainland women in Hospital Authority (HA) hospitals has been relatively steady in recent years. There were 7885 such cases in 2000, 7337 in 2001, 8235 in 2002 and 4214 in the first six months of 2003. Owing to the decline in the overall childbirth rate in Hong Kong, we have actually observed a decline in the total number of childbirths in HA hospitals during the same period. Therefore, the obstetric service of public hospitals has been able to cope with this workload.
>
> (Legislative Council 26 November 2003, 1627)

At this same hearing, the Secretary of Security made the point that "about 84% of the spouses of the women mentioned...are Hong Kong residents", and that these children could obtain the right of abode and settle in Hong Kong regardless of where they were born (ibid.). The fact that early in the process most of the mainland pregnant women coming to give birth were spouses of Hong Kong residents (Table 9.1) helped to reinforce the message that this was a non-threatening process.

Due to rising public and political concerns about mainland maternity migration, the first refinement in the dividing practices was to separate out mainland pregnant women/babies whose spouses/fathers were Hong Kong residents—designated "Type I Non-Eligible Persons (NEPs)"—from mainland pregnant women/babies whose spouses/fathers were not Hong Kong residents—designated "Type II NEPs". Type II NEPs also began to be called "doubly non-permanent resident" pregnant women and babies, which in public discourse was shortened to the "double-nons" or "double-nots",

Table 9.1. Live births in Hong Kong, 2001–2016

Year	Total live births	Live births to mainland women whose spouses are Hong Kong permanent residents	Live births to mainland women whose spouses are not Hong Kong permanent residents[1]	Live births to mainland women, Other[2]	Subtotal
2001	48219	7190	620	NA	7810
2002	48209	7256	1250	NA	8506
2003	46965	7962	2070	96	10128
2004	49796	8896	4102	211	13209
2005	57098	9879	9273	386	19538
2006	65626	9438	16044	650	26132
2007	70875	7989	18816	769	27574
2008	78822	7228	25269	1068	33565
2009	82095	6213	29766	1274	37253
2010	88584	6169	32653	1826	40648
2011	95451	6110	35736	2136	43982
2012	91558	4698	26715	1786	33199
2013	57084	4670	790	37	5497
2014	62305	5179	823	22	6024
2015	59878	4775	775	16	5566
2016	60856	4370	606	3	4979

Source: *Hong Kong Population Projections 2017–2066* (Hong Kong: Census and Statistics Department, 2017), p. 45.

1 Includes Hong Kong non-permanent resident spouses (spouses from the mainland living for less than seven years in Hong Kong).
2 Mainland mothers who did not provide the father's residential status.

reflecting not only their legal non-status but also their status as an abject Other in the eyes of many Hong Kong residents. This dividing practice signals the beginnings of a rise in domopolitics in Hong Kong; it marks out the "right kind" of mainland pregnant women and babies from the "wrong kind", not on the basis of socioeconomic criteria but according to a more exclusionary nationalistic metric that treats birthright citizenship as a right for "properly" Hong Kong people, but as a threat when it applies to the babies of mainland Chinese couples.

In 2004, concerns about overcrowded maternity wards and mainland women leaving without paying their bills were raised at a LegCo hearing with the Secretary of Health, Welfare and Food (Legislative Council 27 October 2004, 711–715). Again, the Secretary was mainly reassuring: "The overall utilization rates of obstetrics and neonatology services in HA (public) hospitals were in the range of 56% to 78% over the past few months and there was no evidence of substantial increase." The only response deemed necessary was to increase staffing, which had declined in recent years due to low birthrates, in order to meet this new demand (ibid., 714).

On 10 November 2004, concerns were once again raised in the LegCo about mainland pregnant women giving birth in hospitals in the northern territories nearest the border, and thereby affecting access to services by locals. The legislators also raised the issue of the large number of mainland pregnant women arriving at HA hospitals after midnight. The Secretary for Health, Welfare and Food responded that utilisation rates had not exceeded capacity. He also stated that 29 per cent of mainland pregnant women arrived between midnight and 6:00 am, somewhat higher than the 23 per cent for Hong Kong pregnant women, and speculated that the fees and way they were charged was probably responsible—HK $3,300/day for NEPs as opposed to HK $100/day for locals—leading more mainland women to wait until after midnight to avoid an extra day's charge (Legislative Council 10 November 2004, 1197). In order to deal with this, the northern hospitals adjusted their staffing, increased training in midwifery, and were monitoring night shift staffing levels to ensure they were adequate to meet these new demands. What the Secretary was mainly concerned about was not the stress placed on public hospitals, but the lack of antenatal checkups by mainland pregnant women in Hong Kong, and the fact that several women left within one day of arrival in order to minimise the cost, which was likely increasing the risk to the women and their newborns. He even noted that on occasion women left their newborns behind if they needed neonatal care, since the babies' hospital fees were set at Hong Kong residents' rates. In order to address these problems, the Secretary recommended changing the fee structure from a daily rate to an average delivery fee, and to increase the deposit required of NEPs (ibid., 1198–2004). These recommendations were implemented in September 2005, when charges for mainland pregnant women were set at HK $20,000, said to reflect the full recovery costs for an average three-day stay for mother and child (Legislative Council 13 June 2005, 3). These adjustments successfully reduced the number of mainland pregnant women giving birth in public hospitals, which never again came close to the 2005 level (Table 9.2).

Through the mid-2000s, then, although local elected officials responding to public concerns and complaints continually raised mainland maternity migration as a problem of biopolitical security, administrative officials in charge of managing unease—the heads of Security; Health, Welfare and Food; and the Hospital Authority—all sought to reassure the politicians and the public, and to present maternity migration as an opportunity which, if managed properly, would benefit Hong Kong. The adjustments made—increased staffing and medical personnel training, changing shift schedules, altering the way fees were charged and collected—all fitted comfortably under a governmental approach that treated mainland maternity migration as good circulation. Still, a domopolitical undercurrent was present in public statements beginning in 2003, and in the dividing practices separating out Type I "acceptable" babies from Type II "undeserving" mothers and babies.

Table 9.2. Live births to mainland women in public and private hospitals, 2002–2015

Year	Total	Public Hospitals	Private Hospitals
2002	8506	8248 (97%)	258 (3%)
2003	10128	8793 (87%)	1335 (13%)
2004	13209	10992 (83%)	2217 (17%)
2005	19538	13911 (71%)	5627 (29%)
2006	26132	12047 (46%)	14085 (54%)
2007	27574	9099 (33%)	18475 (67%)
2008	33565	10741 (32%)	22824 (68%)
2009	37253	10431 (28%)	26822 (72%)
2010	40648	10568 (26%)	30080 (74%)
2011	43982	10556 (24%)	33426 (76%)
2012	33199	3320 (10%)	29879 (90%)
2013	5497	385 (7%)	5112 (93%)
2014	6024	361 (6%)	5663 (94%)
2015	5566	278 (5%)	5288 (95%)

Sources: 2007–2015: *Hong Kong Government News*, 12 July 2017. "LCQ9: Policies and statistics of Mainland residents coming to study, work and settle in Hong Kong," Annex 5; 2002–2006: LC Paper No. CB(2) 1601/06–07(01), 13 April 2007. "Impact of use of obstetric services by Mainland women on public hospital resources," Annex.

The late 2000s: Maternity migration as a problem of security

From 2003 to 2006, the number of live births to mainland women in Hong Kong increased from 10,128 to 26,132 or from 21.6 to 39.8 per cent of the total. Most of this increase was among babies born to women whose spouses were not Hong Kong permanent residents: 2,070 (4.4 per cent) in 2003 to 16,044 (24.4 per cent) in 2006. As this occurred, public, media and political opposition to mainland maternity migration intensified. Misleading assertions that public hospitals were being flooded with "double-nons", displacing native Hong Kong mothers and babies, came to dominate public discourse. Ultimately, healthcare officials within the city administration realigned their position accordingly. By 2007 a domopolitical turn was clearly occurring, which became even more pronounced following the economic reversal of fortunes between Hong Kong and the mainland after 2008. Hong Kong's economy, tied to global markets, was caught up in the recession that rocked the core capitalist economies, while China's relatively insulated economy surged at the same time. Increasingly, mainland maternity migration was framed as a problem of security that must be curtailed.

The year 2007 marks a turning point in the dividing practices associated with maternity migration and corresponds with rising anti-mainland nationalism in Hong Kong. Already at the end of 2006, at a meeting between members of the LegCo and the Kowloon City District, mainland pregnant women were presented as a problem not only for public healthcare services, but as a "problem...that had far-reaching consequences and put tremendous pressure on Hong Kong as a whole" (Legislative Council 9 November 2006, 2). In addition to calling on the Immigration Department to restrict entry and on the HA to force these women to use private hospitals, the Chairman of the Kowloon City District Council called on the government to raise hospital fees, not to cover costs, but "to discourage them from coming to Hong Kong" (ibid., 5). At the same time, members called for "measures to promote child-birth among the people of Hong Kong" (ibid.).

Meetings were held between LegCo members and senior administrators in Health, Welfare and Food and the Hospital Authority on January 8 and 10, 2007. Although a representative of the administration noted that the 2005 reforms had already substantially reduced the use of public hospital services by mainland women, he announced that the HA had nonetheless decided to raise the fees to HK $39,000 for women with a booking (set at the high-end level charged by private hospitals) and HK $48,000 for women without a booking (Legislative Council 8 January 2007, 7–8).

The objectives behind the new restrictions were said to be to: "(1) ensure that local pregnant women are given proper obstetric services and priority to use such services; (2) limit the number of non-local pregnant women coming to Hong Kong to give births to a level that can be supported by our health-care system; and (3) deter dangerous behaviour of non-local pregnant women in seeking emergency hospital admissions through Accident and Emergency Departments shortly before labour" (Progress Report April 2007, 2). To ensure that local expectant mothers received priority, bookings by mainland women would only be accepted if there was excess capacity. The booking system was also presented as a way to ensure antenatal checkups, to lower the risk of disease being spread from mainlanders to local pregnant women and healthcare workers, to restrict numbers and to gauge demand so that the HA could plan for any needed expansion (Legislative Council 8 January 2007, 8–9). To help ensure that the booking system would have the desired effect, the Immigration Department was mandated to check visitors for signs of advanced pregnancy (28 weeks or later), and to demand booking certification. Entry could be denied if the proper paperwork was not produced. The Department of Health provided medical staff to the Immigration Department to help border guards determine the stage of pregnancy. These "border baby patrols" physically examined women they suspected of being past the 28-week cut off "with a tape measure and fundal height chart" (Cheng 2007, 982).

This new system was implemented on 1 February 2007. According to a report released in April 2007, the system was working smoothly: nearly all non-local pregnant women had made bookings, the total number of mainland women giving birth in public hospitals in Hong Kong was down by more than a third compared to the same period in 2006, local births were up slightly, and the Immigration Department conducted 6,698 secondary exam-inations of pregnant women crossing the border, denying entry to 320 (Pro-gress Report April 2007, 4–5).

The new higher fees produced substantial profits for the public hospitals ($600 million from 1 February 2007 to 31 October 2008 alone [*Hong Kong Government News* 10 December 2008]). The number of mainland expectant mothers giving birth in public hospitals decreased further, as more shifted to private hospitals (see Table 9.2 above). For three of the four years between 2008 and 2011, the HA also used the new criteria to suspend bookings for births by mainland women in public hospitals before the end of the year, not because a quota had been met, but in order to ensure adequate obstetric services in public hospitals would be available for local Hong Kong women. In 2011 the suspension was enacted in April, triggering an increase in the number of mainland women using the ER to deliver. Private hospitals also willingly expanded their obstetric services to accommodate this new source of revenue.

If the purpose of the new, more restrictive conditions under which mainland women could give birth in Hong Kong were truly designed to meet the objectives proclaimed, they were unnecessary. As Table 9.2 shows, the 2005 adjustments effectively lowered the number of mainland women giving birth in public hospitals, resolving the first two criteria. Administrative officials themselves noted the success of the 2005 regulations, even as they announced the new, more restrictive conditions in 2007. The third criterion—to prevent mainland women from engaging in the dangerous behaviour of rushing to the ER at the last moment to deliver their babies without a booking—was made worse under the new policy. The 2007 restrictions failed to do the one thing that Hong Kong nationalists were increasingly demanding, and that was to reduce the number of mainland couples giving birth to mainland babies who would have the right of abode in Hong Kong.

Government officials next sought to curtail the expansion of obstetric services in private hospitals. The Secretary for Food and Health presented this as necessary due to the loss of doctors and nurses from public hospitals who were being lured away to work in the more lucrative private sector. As he put it at a meeting with private hospital managers in 2011, "The situation in public hospitals is particularly serious. The Hospital Authority (HA) has seen an increase in the turnover of obstetric doctors, nurses and midwives due to the booming private obstetric market" (*Hong Kong Government News* 4 April 2011; see also *Shenzhen Economic Daily* 6 April 2011). Following this meeting, private hospitals agreed to cap the number of mainland women giving birth.

Despite the fact that the regulations put in place were working to reduce the number of mainland women giving birth in public hospitals, public organisations, media reports and politicians in Hong Kong's LegCo continually referenced the total number of births to mainland women, creating the misleading perception that they were "flooding" or "swarming" the maternity wards of public hospitals, overwhelming the staff and facilities and displacing the truly deserving native Hong Kong expectant mothers. For example, a reporter covering a march by local Hong Kong pregnant women and young couples against mainland maternity migration wrote the following:

> It is a reality in Hong Kong, as mainland parents without proper local connections are flooding to the city to give birth. It may improve Hong Kong's low birth rate, but since the number of mainland mothers is climbing faster in recent years, the demands on our medical services have reached critical proportions…And many more will be coming. With a 1.4 billion population on the mainland, it is estimated there will be about 20 million babies born annually. Even if one percent of pregnant mainland women come here to give birth, it will render our medical services inoperable (Sio-Chong 2012).

This imaginative geography of mainlanders without "proper local connections" threatening the future of "native" Hong Kongers targeted the "doubly non-permanent resident pregnant women" and their "alien" babies who would become Hong Kong permanent residents at the moment of birth.[6] The affective message that this flood could not be accommodated and must be stopped became irresistible, and any government official arguing that mainland maternity migration was a good thing for Hong Kong was quickly drowned out or targeted by the increasingly assertive anti-mainland activist groups. NGOs such as "It's Time for Us To Say NO!" mobilised opposition to mainland maternity migration online (www.facebook.com/itstimetosayno), in the press with ads and in the streets with protest rallies. According to the administrator of this Facebook group (interview conducted in March 2013),

> The government saw citizens with no political background—pregnant women and children—protesting in the rain, and realized the seriousness of the problem. They took action against the mainland families coming to Hong Kong, such as Mr. Leung's administrative measure on no mainland family giving birth in Hong Kong…The ultimate goal of this community is to stop and to cancel all the rights of abode in Hong Kong of anchor children…No anchor children is the consensus of Hong Kong society.

The rising public hostility toward mainland mothers and their "anchor babies" fed into the geographical imagination of mainland Chinese as locusts poised to swarm across the border and strip Hong Kong bare.[7] As April Zhang (2012) put it, "Protests against pregnant mainlanders were because locals felt entitled to maternity beds and mainland mothers were taking away this privilege…Hong Kong people are now struggling with the feeling of being both 'colonized' by mainland China and invaded by mainlanders, whether they be tourists, mothers-to-be or drivers". The imagery of rich mainlanders coming to occupy space in maternity wards was added to those coming to buy luxury goods, property and baby formula, and whose children were coming to take up seats in classrooms meant for native Hong Kong children. These anti-mainland images merged with the imagineering of mainland Chinese as uncouth, speaking loudly in public, eating in improper places and urinating and defecating in public spaces.[8] Anti-mainland maternity migration groups also joined with the rising anti-Beijing, pro-democracy/pro-autonomy protests movements that became much more visible and assertive at this time.

In 2011, the government set a quota of 34,500 births to mainland pregnant women for 2012–3,400 for public hospitals and 31,000 for private. After the increasing number of mainland women using ERs for delivery in 2011, the HA raised the fee level for deliveries without bookings from HK $48,000 to $90,000 in 2012 and increased border surveillance (Hospital Authority 11 May 2012). Finally, the new Chief Executive-elect, CY Leung, announced a "zero quota" for 2013 until the broad societal impact of mainland women giving birth in Hong Kong could be assessed (Luk 2012; Tsang 2012). Leung, who was selected by Beijing and deeply unpopular among the more nationalistic Hong Kong public, was seeking to gain favour within the growing anti-mainland community with this move.[9] The ban explicitly targeted mainland couples seeking to give birth in Hong Kong; mainland pregnant women with Hong Kong spouses were exempted from the ban, though even they were not allowed to make use of public hospital space. This exemption laid bare what the biopolitical unease was actually about: the mainland Chinese-ness of the babies automatically receiving the right of abode represented for many Hong Kong nationalists an alien foetal invasion, or a surreptitious takeover of Hong Kong by the mainland, subverting the promise of autonomy made under the one country, two systems policy.

Following the ban, the managers of biopolitical unease redirected their attention in two ways. First, security concerns shifted from the macro- to the micro-level. Reporting now focused on individual efforts by mainland women to evade the ban: of "fake" marriages to Hong Kong husbands and "bogus" female students admitted to Hong Kong universities in order to give birth in Hong Kong (e.g., Immigration Department Review 2012, 8 February 2013; Cheung 2013; *Hong Kong Government News* 30 January 2013a; 30 January 2013b; 1 February 2013; *Sing Tao Daily* 15 January 2013), of agents who help sneak pregnant women across the border so they can "gate-crash" the Emergency Rooms of public hospitals, and of property owners arrested for running guest houses for mainland pregnant women. These stories appeared with increasing frequency in 2012 and came to dominate the news associated with mainland maternity migration in 2013, with monthly news reports of ER births to mainland women (e.g., *Hong Kong Government News* 5 December 2014). These reports seemed designed both to reassure the public (there are fewer cases/we are stopping them) and to elevate public anxiety (they continue to break through our defences/they have greedy, traitorous helpers on the inside).

The penalties for violating the ban became severe, with mainland pregnant women given prison sentences of several months to several years (e.g., Hong

Kong Government News 1 February 2013, 11 February 2014). Mainland maternity migration is now presented as extremely bad circulation; mainland pregnant women are portrayed as criminals looking for any and every opportunity to slip through Hong Kong's defences and give birth to yet another "alien" citizen. Officials now seek to reassure the public that they are sparing no expense to prevent even one mainland pregnant woman from succeeding. With this domopolitical turn, Hong Kong's managers of biopolitical risk have rejected the normalising regulatory mechanisms of governmentality in favour of the sovereign prohibition.

The second shift was temporal. Rather than focus unease on the present (mainlanders poised at the border NOW), fear and anxiety were redirected toward the future and past, toward the large number of Hong Kong born school-aged children in the mainland who will soon enter Hong Kong to make use of education, healthcare, welfare and other social services. By 2012, the phrase "the longer-term implications of Type II babies on education, social and other related services" (Steering Committee on Population Policy 2012, iii) became a dominant rationale for ending mainland maternity migration and for voicing concerns about how to deal with the babies already born. So, even while the Steering Committee announced that the population was aging more rapidly than earlier projections predicted and that Hong Kong was in desperate need for increasing the birthrate to spur population growth, they also proclaimed that "Type II children are not a solution to our demographic challenge" (*Hong Kong Government News* 25 October 2013). At the same time, the government announced a number of "enhanced measures...to build human capital" in Hong Kong, including adjustments "to attract more foreign talents with international work experience" or higher education, making it easier for them to enter and stay (*China Briefing* 7 May 2015).

At the end of 2012, Hong Kong's Secretary of Justice sought to reverse the 2001 court ruling that provided mainland Chinese babies born in Hong Kong the right of abode (*China Daily* 15 September 2012). Anti-mainland activists have also increasingly called for the removal of the right of abode for children born to mainland couples—a form of "backward uncitizening"—as the only solution to this past-future problem. If a referendum were held to deprive the children of mainland parents of their right of abode in Hong Kong, there is little doubt it would pass by a wide margin.

Conclusions

Mainland maternity migration to Hong Kong, which by all accounts should have been perceived as "good circulation", quickly produced affects of fear and loathing among the Hong Kong public and contributed to the performative materialisation of mainland Chinese-ness as the constitutive outside and abject other of Hong Kongers' increasingly exclusionary and nationalistic enactment of place-identity. By the end of the 2000s, each negative public incident—a mainland child peeing in public, a mainland adult eating on the subway, a mainland pregnant woman rushing to the ER to deliver a baby without a booking, cross-border traders buying up baby formula, a local child having to travel a greater distance to school to accommodate a "cross-border student"—was stitched together to produce a border exteriorising mainland Chinese not only as Hong Kong's "not us", the constitutive outside, but as not properly human, as swarming locusts stripping Hong Kong bare and leaving nothing for its "native" people. In the popular imaginary, spitting, urinating, defecating, spreading disease and giving birth all came to be connected as bodily excretions through which mainland Chinese were threatening the very existence of Hong Kong and its people. The accompanying rise in anti-mainland nationalism has also raised serious doubts about the potential for successful integration of Hong Kong under the "one country, two systems" policy.

The case of mainland maternity migration to Hong Kong also helps to demonstrate the utility and versatility of domopolitics and the governmentality of unease. These extensions of Foucault's lectures on governmentality and biopolitics provide indispensable tools with which to interrogate the dividing practices that determine how "good" and "bad" circulation are produced. This chapter shows how they can be put to work in more performative research on nationalism and homeland politics that is not state-centric, does not start from a top-down or centre-out approach to the politics of affect (in this case the management of unease), and that goes beyond the more narrowly cast focus on dividing practices that materialise citizen and not-citizen in the context of asylum-seekers. In particular, this work emphasises the need to shift attention away from citizenship/non-citizenship dividing practices and toward those that materialise homeland nation/Others within—not only non-status Others but also those with citizenship. Using domopolitics and the governmentality of unease to interrogate exclusionary practices against maternity migration and efforts to roll back birthright citizenship is both timely and critical, given the ways in which the far right has used these as wedge issues in their recent rise.

Finally, this chapter highlights the importance of reproduction in the study of biopolitics, something that has received surprisingly little attention to date. While Mills (2017) has drawn attention to the centrality of reproduction "to the operation of biopolitics", and illustrates this with the issues of birth control, prenatal testing and abortion politics, the biopolitical management of life associated with maternity migration, birthright citizenship and homeland nationalism remains largely unexplored. This work represents a preliminary effort to begin addressing this critical gap.

Notes

1 This resulted from the 2001 Hong Kong Court of Final Appeal decision in the case of the Director of Immigration v. CHONG Fung Yuen, which ruled that "Chinese

citizens born in Hong Kong before or after the establishment of the Hong Kong Special Administrative Region (HKSAR) are permanent residents of the HKSAR upon whom Article 24(3) confers the right of abode, regardless of the immigration status of their parents" (Legislative Council Paper No. CB(2)1748/04–05(03), 13 June 2005).

2   This was supposed to be a temporary expedient while the government studied the social, economic and cultural impacts of maternity migration on Hong Kong, but as of May 2018 the ban has not been lifted.

3   These dividing practices, and particularly those that produce nationalised and racialised borderings of populations, are the focus of *Society Must Be Defended* (Foucault 2003).

4   Though she does not use the framework of domopolitics, Eithne Luibhéid (2013) provides a brilliant analysis of the way in which childbirth among asylum seekers in Ireland was used to construct them as illegal immigrants and ultimately to revoke birthright citizenship.

5   Beginning on 1 January 2016, China loosened restrictions to allow two children per couple.

6   This discourse of pregnant migrants with no proper local connections as a biopolitical threat to legitimate native people was also found in Ireland in the run-up to the referendum restricting birthright citizenship in 2004 (Luibhéid 2013).

7   Funds were raised from small online donations within a week to pay for a full page ad in *Apple Daily*, a popular Hong Kong newspaper, with a headline "Hong Kong people, we have endured enough in silence," and depicting the mainland Chinese as a giant locust overlooking the skyline of Hong Kong. This made international news – see, for example, Lim 2012; South China Morning Post, 1 February 2012.

8   These negative stereotypes of mainland Chinese immigrants held by Hong Kongers are not new. See, for example, Law and Lee (2006).

9   Leung followed this with a series of policies designed to curry favour with the more anti-mainland segment of Hong Kong's electorate: in particular, decisions to limit the amount of baby formula that mainlanders could carry across the border to two cans and to charge a 15 per cent tax on foreigners who wished to buy property in Hong Kong (Lee and He 2013).

Acknowledgements

I would like to gratefully acknowledge the support I received for this research project from a variety of sources: the Leon D. Epstein Faculty Fellowship from the University of Wisconsin—Madison, the University of Shenzhen, and especially the assistance of Professor Ding Wei and her graduate students, the University Services Center for China Studies at the Chinese University of Hong Kong (Professor Chan, Director), and the David C. Lam Institute for East–West Studies at the Hong Kong Baptist University (Professor Si-ming Li, Director).

References

Arendt, Hannah. 1951. *The Origins of Totalitarianism*. New York: Harcourt Brace.
Berlant, L. 1994. "America, 'Fat', the Fetus." *Boundary 2 21(3)*: 145–195.
Bigo, Didier. 2002. "Security and Immigration: Toward a Critique of the Governmentality of Unease." *Alternatives 27*: 63–92.

Brown, Wendy. 2010. *Walled States, Waning Sovereignty.* Cambridge: MIT Press.

Census and Statistics Department. 2011. *Hong Kong Monthly Digest of Statistics.* Featured Article: "Babies born in Hong Kong to Mainland Women." September.

Cheng, Margaret Harris. 2007. "Hong Kong Attempts to Reduce Influx of Pregnant Chinese." *The Lancet*, vol *369* (March 24): 981–982.

Cheung, S. 2013. "Women in Marriage Scams Give Birth in HK," *South China Morning Post*, 9 February.

China Briefing. 7 May 2015. "Hong Kong Announces Enhancement Measures for Immigration Policy."

China Daily (Hong Kong Edition). 15 September 2012. "Cure for 'Non-local' Pain." HK Opinion.

Cisneros, Natalie. 2013. "'Alien' Sexuality: Race, Maternity, Citizenship." *Hypatia: A Journal of Feminist Philosophy 28(2)*: 290–306.

Connolly, William. 1995. *The Ethos of Pluralization.* Minneapolis: University of Minnesota Press.

Darling, Jonathan. 2011. "Domopolitics, Governmentality and the Regulation of Asylum Accommodation." *Political Geography 30*: 263–271.

Darling, Jonathan. 2014. "Asylum and the Post-political: Domopolitics, Depoliticisation and Acts of Citizenship." *Antipode 46(1)*: 72–91.

Farley, Robert. 2015. "Trump Challenges Birthright Citizenship" *FactCheck.org* Accessed 22 September 2018.

Foucault, Michel. 2007. *Security, Territory, Population.* Lectures at the College de France 1977–1978. New York: Picador.

Foucault, Michel. 2003. "Society Must Be Defended": Lectures at the Collège de France, 1975–1976 (Vol. 1). New York: Macmillan.

Hong Kong Government News. 5 December 2014. "29 A&E Birth Cases in November."

Hong Kong Government News. 11 February 2014. "3 Mainlanders Jailed for Illicit Births."

Hong Kong Government News. 25 October 2013. "Population Policy-Thoughts for HK."

Hong Kong Government News. 1 February 2013. "Mainland Pregnant Woman Jailed for Making False Representation to Immigration Officer and Attempting to Obtain Services by Deception."

Hong Kong Government News. 30 January 2013a. "Warning Issued to Pregnant 'Students'."

Hong Kong Government News. 30 January 2013b. "LCQ3: Mainland Pregnant Women giving Birth in Hong Kong."

Hong Kong Government News. 2 February 2012. "LCQ8: Services of Maternal and Child Health Centres."

Hong Kong Government News. 11 May 2011. "LCQ4: Obstetric Services."

Hong Kong Government News. 4 April 2011. "SFH meets Private Hospital Operators."

Hong Kong Government News. 10 December 2008. "LCQ14: Non-local Pregnant Women using HA's Obstetric Services."

Hospital Authority. 2012. "New Obstetric Fee for Non-Eligible Persons Gazetted Today." 11 May. Press Release.

Hynek, Nik. 2012. "The Domopolitics of Japanese Human Security." *Security Dialogue 43(2)*: 119–137.

Immigration Department Review 2012 (Summary). 8 February 2013. *Hong Kong Government News.*

Ingram, Alan. 2008. "Domopolitics and Disease: HIV/AIDS, Immigration, and Asylum in the UK." *Environment and Planning D: Society and Space 26*: 875–894.

Law, Kam-yee and Lee, Kim-ming. 2006. "Citizenship, Economic and Social Exclusion of Mainland Chinese in Hong Kong." *Journal of Contemporary Asia 36(2)*: 217–242.

Lee, Colleen and He, Huifeng. 2013. "Milk-powder Row Symptom of a Wider Rift with Mainland." *South China Morning Post*, 17 March.

Legislative Council. 8 January 2007. "Panel on Health Services: Minutes of Meeting." LC Paper No. CB(2)*1043/06–07*.

Legislative Council. 9 November 2006. "Giving Due Regard to the Issue of Mainland Pregnant Women Giving Birth in Hong Kong." LC Paper No. CB(2)*1225/06–07(01)*.

Legislative Council. 13 June 2005. "Legislative Council Panel on Health Services: Hospital Fees and Charges-Non-eligible Persons and Private Patients." LC Paper No. CB(2)*1748/04–05(03)*.

Legislative Council. 10 November 2004. "Mainland Pregnant Women Giving Birth in Hong Kong." *Official Record of Proceedings*, pp. 1196–1204.

Legislative Council. 27 October 2004. "Mainland Women Giving Birth in Hong Kong." *Official Record of Proceedings*, pp. 711–715.

Legislative Council. 26 November 2003. "Overstaying Mainland Women Giving Birth in Hong Kong." *Official Record of Proceedings*, pp. 1627–1628.

Lim, Louisa. 2012. "For Hong Kong and Mainland, distrust only grows." *NPR All Things Considered*, 23 March.

Luk, E. 2012. "Door Shuts on Moms: Leung Shocks Hospitals with 'Zero Quota'." *The Standard*, 17 April, p. 1.

Luibhéid, Eithne. 2013. *Pregnant on Arrival: Making the Illegal Immigrant*. Minneapolis: University of Minnesota Press.

Mancini, JoAnne and Finlay, Graham. 2008. "'Citizenship Matters': Lessons from the Irish Citizenship Referendum." *American Quarterly 60(3)*: 575–599.

Mills, Catherine. 2017. "Biopolitics and Human Reproduction" In *The Routledge Handbook of Biopolitics* edited by Sergei Prozorov and Simona Rentea, 281–294. Abingdon: Routledge.

Ngai, Mae M. 2007. "Birthright Citizenship and the Alien Citizen." *Fordham Law Review 75(5)*: 2521–2530.

Nyers, Peter. 2006. "Accidental Citizenship: Acts of Sovereignty and (Un)making Citizenship." *Economy and Society 35(1)*: 22–41.

Progress Report. April 2007. "Motion on 'Non-local Pregnant Women Giving Birth in Hong Kong' Legislative Council Meeting on 10 January 2007." Hong Kong: Health, Welfare and Food Bureau and Security Bureau.

Shenzhen Economic Daily. 6 April 2011. "The Situation of Gynecologist Drain in Hong Kong Public Hospitals is Serious."

Sing Tao Daily. 15 January 2013. "Adopt Piecemeal Measures to Properly Block Double-not Babies."

Sio-Chong, S. 2013. "Who Causes the Hospital Beds Shortage for Local Mothers?" *China Daily* (Hong Kong Edition), 19 January, HK Opinion.

Somers, Margaret R. 2008. *Genealogies of Citizenship*. Cambridge: Cambridge University Press.

Steering Committee on Population Policy. 2012. *Progress Report*. Hong Kong: Census and Statistics Department, May.

Task Force on Population Policy. 2003. *Report of the Task Force on Population Policy. Summary of Recommendations.* Hong Kong.

Titley, Gavan. 2012. "Getting Integration Right? Media Transnationalism and Domopolitics in Ireland." *Ethnic and Racial Studies 35(5)*: 817–833.

Tsang, Emily. 2012. "Zero Quota May Ease Staffing Crisis." *South China Morning Post*, April 22.

Walters, William. 2004. "Secure Borders, Safe Havens, Domopolitics." *Citizenship Studies 8(3)*: 237–260.

Wang, Sean. 2017. "Fetal Citizens? Birthright Citizenship, Reproductive Futurism, and the 'Panic' over Chinese Birth Tourism in Southern California." *Environment and Planning D: Society and Space 35(2)*: 263–280.

Zhang, April. 2012. "Hong Kong Identity Caught between Reality and Insecurity." *South China Morning Post*, 17 October.

10 National pasts and biopolitical futures in Serbia

Carl T. Dahlman

Introduction

Faced with what it calls "the white plague" of low birthrates, the Serbian government's Council of Population Policy and Ministry of Culture in 2018 announced a new media campaign to increase the Serb population. Among the winning slogans were rhymes "Ljubav i beba – prvo što nam treba" (Love and baby – the first thing we need!), imperatives "Rađaj, ne odgađaj!" (Give birth, don't delay!) and the religiously toned "Čuda se ne dešavaju. Čuda se rađaju" (Miracles don't happen. Miracles are born) (*Dnevnik* 2018). Not everyone approved. A deputy prime minister charged with gender equity issues said the slogans "publicly shame women" (*B92* 2018). Feminist groups decried the government's failure to address the terrible hardships that motherhood creates in Serbia's anaemic economy, echoing the complaints of the Women in Black and other activists since the 1990s (Papić 2010; e.g. Zajovic 1995).[1] Still, many popular media outlets and public voices embraced the government's campaign as a worthy response to the fact that Serbian birthrates have been falling for decades.[2]

Pronatalist campaigns like Serbia's reveal how reproduction figures in a wider geopolitical imagination of national survival. In this chapter, I approach the geopolitics of reproduction through the development of biopolitical knowledge and interventions as a constitutive aspect of modern nationalism. These biopolitical practices, traced here through evolving state epistemologies that make populations calculable and bodies legible for intervention, are often missing from geopolitical analyses of nationalist movements, which primarily focus on territory (but compare Anderson 2006). I argue that biopolitics and territorial politics are necessary and complementary parts of any nationalist movement as it captures the state and configures life in assurance of a national future.

Of course, the politics of life can be deadly. Serbia's nationalist leader of the 1990s, Slobodan Milošević, drew on long-standing nationalist claims that Kosovo and other parts of Yugoslavia were part of a Serbian homeland taken from them by other groups. Speaking to a rally of angry Kosovar Serbs in 1987, Milošević invoked an imagination of reproductive continuity that linked his audience to mythic Serb nationalist claims to the province: "You should stay here because of your ancestors and because of your descendants. Otherwise you would disgrace your ancestors and disappoint your descendants" (FBIS Daily Report 27 April 1987). The ideology of national homeland drove Serbia's expansionist politics that led its forces to ethnically cleanse Albanians and others from territory they conquered during the war of the 1990s. Territorial conquest, however, not only makes historical claims to land, it also occupies space for some future nation yet to be born.

As this chapter explains, the contemporary imagination of reproductive continuity seeks to validate Serbian national claims to territory in a multi-national space. In the 150 years since its independence, Serbia has been governed by monarchs, fascists, communists, authoritarian nationalists and illiberal democrats. Nonetheless, a biopolitical vision of population and territory has developed alongside and in spite of those ideological shifts. Serbian nationalists have long argued that their cultural tenacity was proven by population growth that justified territorial sovereignty. The actual decline in Serbian fertility and population size therefore produced anxiety of becoming a minority in someone else's country. Yet Serbia itself has always been a multi-national territory, a reality that Serbian nationalists still view as a threat to the future of their people's homeland.

This chapter also presumes that Serbia's biopolitical nationalism is just a specific example of the more general problematic about how states come to know and manage populations (Bashford 2016). Biopolitics, as Michel Foucault and his interpreters explain, is a process of making people and territory legible and calculable and thereby available for management, intervention and control by the state and para-statal institutions (Lemke 2001, 2011; Legg 2005). State goals for population growth and requisite reproductive practices, moreover, are exceptionally gendered techniques that attempt to make women governable targets of biopolitical programmes. At the same time, nationalism hides women behind institutions of family and traditional community where their reproductive labour and risks are concealed by conservative social and cultural practices (Deutscher 2008).

Biopolitics, the problematic occurring at the boundary between state institutions and the people, comprises multifarious practices around vitality and survival. These practices are intended to assure the emergence of biopower in a population whose growth must be managed to secure the future. Women's bodies, rendered as sites of both reproduction and property, become targets of nationalism's patriarchal concern with managing the conditions of future life, even as it undermines the lives of women themselves (Mostov 2000; Papić 2010). Serbia's biopolitical interventions into the field of reproduction, such as pronatalism, were also part of a wider geopolitical competition for territory and an attendant anxiety of reproductive outpacing by other groups, a focus on the bodies of the Other (Žarkov 2007). Nationalism, therefore, is as much about managing the arrival of a future population as it is about controlling a territory today.

Territory, population and future nations

Geopolitical interpretations of Serb nationalism commonly focus on the spatial homology of state authority, national population and territorial claims; in other words, the territoriality of the nation-state (Savić 2014). Nationalism, as a category of practice, relies heavily on historical claims to cultural identity and territory. As a category of analysis, nationalities studies have tended to adopt this rear-facing perspective. Yet nationalism is also a forward-facing politics that addresses the most pressing question, "What is our future?" Theories of nationalism commonly focus on the rather masculinist public histories of how territory is conquered and defended and how material wealth is produced, ignoring the social reproduction through fertility, childrearing and enculturation typically assigned to women in the private sphere. In these androcentric explanations of nationalism, as in nationalism itself, women and minorities are erased from historical view and political life yet remain seen as biopolitical targets of state manipulation and direct violence to secure a national future.

Approaching nationalist movements from a biopolitical framework, by contrast, more clearly reveals that gendered constructions of biological and social reproduction are central aspects of nationalism because they are the central creative force that secures the future. Biopolitical explanations necessarily comprise the institutions of social reproduction, especially in the medical and religious fields that transfer patriarchal relations belonging to the household and community into those of law, authority and expertise belonging to the nation-state. The biopolitical *episteme* has fashioned ways to know and calculate population, among them relying on size and growth as indexes of national strength.

As an ideology, moreover, nationalism is keenly interested in "what ought to be" and demography responded with techniques for calculating population trends, allowing the state to imagine and perhaps shape the future's arrival. Birth and death, morbidity, migration, literacy and income became objects of biopolitical knowledge as new political, economic and medical ideas changed the way states related to their societies. Among these techniques, fertility rates, birth registers and life tables emerged as useful measures for states to look into their demographic future and seek control of reproduction as the source of biopower. Thereby the state could reduce its uncertainty regarding the future by intervening into the field of reproduction in pursuit of the biological maxim: "strength in numbers". After all, asks nationalism, "what good is territory if it is empty?"

Biopolitics as historical inquiry

The development of biopolitics as a general concept has opened new lines of historical inquiry about how state epistemologies and techniques created an object called "population" but such studies are less common in the case of specific state practices, such as Serbia (Agamben 1998; Foucault 2003, see also 2007; Rose 2007; Rose and Miller 2010). The development and diffusion of biopolitics certainly accompanied the emergence of scientific fields to aid the state in organising people and territory beginning in the late eighteenth century, made legible in the census, vital statistics, cadasters and maps (Porter 1986). These developments are largely understood as expressions of Enlight-enment rationality culminating in the specific context of nineteenth-century Western European liberalism. Stuart Elden (2007, 573), for example, defines biopolitics as "the means by which the group of living beings understood as a population is measured in order to be governed, and tied to the political rationality of liberalism". How then do we relate the general problematic of biopolitics to a specific history of nationalism during the volatile political transformation of a country like Serbia over the last two centuries?

Foucault (2003, 261) undoubtedly thought biopolitics transcended liber-alism, fascism and socialism as that which is "the essential function of society or the State, or what it is that must replace the State, is to take control of life, to manage it, to compensate for its aleatory nature". This is not to say that biopolitics are non-ideological but rather result from an interface between humanity's political and biological existence, a contested terrain with no single logic. Both capitalist and socialist societies were pri-marily consumed with capturing biopower and managing lives towards cer-tain productive ends. Nationalism thus appears not as a contradiction to these programmes but as an expression of belonging and rights that fore-grounded imagined kinship and descent as the basis of political community. Biological reproduction thus became the site of political struggle to define the qualities of membership and the nation. In turn, these characteristics became the target of interventions that regulate and condition the life of the nation within its territory, generating a geopolitical programme of biological and spatial maximisation (Lemke 2011).

The persistence of nineteenth-century biopolitics in the programmes of socialist societies during the last century reveals the persistence of population as a central problematic for modern states. Foucault's comparison of state racism under the Nazis to the Soviet's desire for a pure classless society through the New Soviet Man was troubled by a lack of historical and poli-tical specificity (Lemke 2011). More recent scholarship, however, reveals how socialist states inherited biopolitical concepts and techniques. Sergei Prozor-ov's (2013) study of Stalinist biopolitics approaches the role of biopolitics under differing ideologies by identifying how states problematise life as an object of revolutionary politics. While all biopolitics result from the imma-nence of governmentality in the problematisation of life, Soviet biopolitics engaged death as part of its revolutionary intervention to "abolish old forms of life and create new ones" (Prozorov 2013, 211). Where lives are unma-nageable by the state, argues Prozorov, the result is a violent, revolutionary expression of power, whether in Ukranian villages during Stalinism or in Iraq under occupation.

Socialist states have also used less violent biopolitical techniques to contend with unruly populations. Recent scholars of socialism and post-socialism have identified alternative interventions meant to manage seemingly ungovernable bodies. In her study of Romanian pro-natalism under Ceausescu, Gail Kligman (1998, 11) highlights the mechanisms of state coercion and bodily resistance among women who were made accountable for reproducing the nation and the "new socialist person". In the wake of the Chernobyl disaster, Adriana Petryna (2013, 12–13) observed how the Ukrainian state and medical authorities generated a form of "biological citizenship" that assigns rights based on a bureaucratisation of contaminated bodies.

Returning our attention to the Serbian state, including the period of Yugoslav socialism, the reproduction of the Serbian nation hinged on what Turner terms "reproductive citizenship", whereby one's political rights and access to resources are shaped by the state's desire to attain numerical advantage for one nation over another (Richardson and Turner 2001; Turner 2003, 2008; but compare Richardson 2017). Even in Western liberal states, reproductive citizenship rewards certain groups for higher birthrates as desirable forms of social and political participation while excluding "undesirable" fertility. In the context of national competition, such as Serbia, reproductive citizenship produces pernicious forms of gender discrimination because institutionalised pronatalism within one group becomes antinatalist discrimination against others. We will see these expressions of biopolitics throughout the following sections, tracing the historical development of Serbian biopolitical and territorial geopolitics from the liberal nationalist period of the nineteenth century to the programmes of ethnic cleansing that attempted to claim an exclusive national homeland in the 1990s.

Territorial nationalism

Modern Serbian nationalism emerged in the nineteenth century, defining itself primarily in opposition to the Ottoman Empire, from which Serbia emerged as an autonomous principality. As an expression of cultural unity, its central institution was the Serbian Orthodox Church, which maintained patriarchal themes in keeping with traditional rural extended families, or *zadruga*, that comprised the imagined hearth of the Serb people during the centuries of Ottoman rule (Halpern and Kerewsky-Halpern 1972). The early principality was largely consumed with securing its independence from the Ottomans and annexing the Serb irredenta beyond its borders, both of which required revenue and soldiers (Vuletić 2012). Serbian officers sent to collect taxes and conscripts reported a countryside where illiteracy and immiseration were near total (Palairet 1995). The conservative peasantry resisted the principality's efforts at centralised authority and increased taxation for much of the century (Stokes 1990).

Serbia's population, in the biopolitical sense of a calculated people, was largely unknown to the prince's government and statistical forays into its midst were rudimentary. The formal grant of autonomy in 1833 included new territory, prompting Serbia to undertake its first tax and military registration in 1834 (Shaw 1978). Serbian Prince Miloš sought to replicate the practices of "civilized countries" that knew their "total and wealth", making registration compulsory and placing vital recordkeeping in the hands of the church

(Vuletić 2012, 3). Regular tax lists were common, as well, particularly as the state spent increasing sums on wars against its neighbours. Biopolitically, the young Serbian state knew its "total and wealth" as men and land. Women, by contrast, were not recorded by name nor age, but only as "total women" under a man's roof (Cvijetić 1984).

The Serbian nation, understood as people of a common tongue and Christian Orthodoxy, was of interest to the Serbian principality largely for the lands they occupied in the wider region. The founding programmes of Serbian nationalism emphasised state expansion into Serb territories liberated from Ottoman rule. Ilija Garašanin's plan (*načertanije*) of 1844 envisioned a unification of Serbs reaching into today's Bosnia, Kosovo and Albania (Hehn 1975). Serbian expansion did not exclude the possibility of a union with other South Slavic peoples but Albanians are not Slavic (Dragović-Soso 2004). Infiltrating Serbian agents were to establish alliances and forestall resistance among local Albanian tribes. Orientalist designs were laid to render Albanians and other minorities "complaisant" with limited religious freedom and token recognition of traditional leaders (Hehn 1975, 165). The expanding Serbian principality would dismantle the Ottoman state "stone by stone" in Garašanin's plan because national territory was destiny:

> The geographic position of the country, its topography, abundance of natural resources, the martial spirit of its inhabitants, their elevated and fiery national feeling, and linguistic and ethnic homogeneity—all contribute to a sense of permanency and a promising future.
>
> (Hehn 1975, 160, 163)

Expansionism and irredentist nationalism became persistent elements of Serbian foreign policy during the nineteenth century. Serbia more than tripled in size as the great powers remapped the Balkans after the Russo-Turkish War (1877–1878), which granted Serbia its independence and the region of Niš, and again after the First and Second Balkan Wars (1912–1913), which awarded Kosovo and Macedonia to Serbia. What Serbia found in its new territories was a more culturally heterogeneous population mired in misery, land poor and illiterate. These conditions in the new territories prompted the newly independent state to take up active interventions into the lives of its population, including the public health concerns of its sickly peoples. Special enumerations of the new territories were conducted in 1879 and of the whole country in 1884. Regular semi-decennial censuses fueled a bureaucratic industry that put out statistical yearbooks, giving statesmen a masterly view of the people and the land.

Making population legible

If we rethink population records and censuses as a distribution of the sensible—the social facts that are visible to the state through data and tables—

then three kinds of populating bodies became apparent to the heterogeneous Serbian state in the late nineteenth century (Rancière 2004). First were the bodies of men who could fight, especially after the crushing defeat against the Ottomans in the 1870s that led to the Treaty of Berlin and Serbia's expansionistic geopolitics. The second kind of populating body was the taxable subject of the agrarian economy, measured at first not as embodied labour but as what they farmed. The third sort of populating bodies were national ones, Serbs in the cultural sense defined primarily by their South Slav language and Orthodox Christianity. In contrast were the Albanians, as well as Catholic and Muslim Slavs, among other linguistic and religious minorities.[3]

Women, however, as social facts of the state for most of the nineteenth century, did not exist as individuals except in ecclesiastic vital records. Early tax rolls and censuses collected the names of men and types of land use; women's names were not recorded until 1862 along with children and their ages. This was significant because the people became more than a momentary source of taxes and conscripts. A population was coming into focus with a specific structure of sex and age; growth could now be estimated—calculation revealed its demographic future. This was the age of "political arithmetic", which tutored statesmen in a descriptive numeracy and basic calculations of wealth and strength (Porter 1986).

Birthrates and growth trends were not the same thing as calculating fertility, however. The field of demographic statistics, which blossomed in Western Europe and replaced political arithmetic, arrived slowly in places like the Balkans, still lacking the bureaucratic capacity for efficient enumeration and record keeping. Making populations legible was less a calculation problem as it was of accurate record keeping and retrieval, stable locational referents and efficient computational resources. It also required a re-conceptualisation of population growth as a calculation of reproduction, shifting the focus from net births to a more intimate measure of women's fertility practices from menarche to menopause. Fertility statistics, such as average number of births, maternal age, stillbirths and illegitimate births, gave the state a sense of "typical" reproductive practices, generating new sites of intervention into the lives of actual women. Yet these calculations were logistically impossible for most nineteenth-century censuses since they required the collection of reliable vital statistics on mothers and births (Ryder et al. 1971). In Serbia, this would have required time-consuming and expensive tabulation and calculation of both census records and ecclesiastic records.

The biopolitical techniques of countries like Serbia continued to calculate fertility from summary statistics; probabilistic models of growth were not calculable until the mid-twentieth century (Zarkovic 1955). The concept of fertility rates, however, were easily imagined in aggregate, national terms. The medical authority emerging alongside demography in the biopolitical state already described a Serbian nation in dire condition, a public health crisis that threatened the nation's future (Stokes 1990). While statistical techniques lagged behind Western Europe, Serbia could already imagine interventions into the lives of its citizens that would improve the conditions for life and the accumulation of national biopower.

Populating a national homeland

As the Serbian state grew in size, its subjects included larger numbers of non-Serbs. In response, it began from 1866 to collect statistics that could identify "national bodies" primarily through language and religious identification. Population became visible as statistical structures of ethno-national difference at each administrative level, producing minorities in one region and majorities in another. These statistical bones were given flesh by geographers and ethnographers, who began to study the "national character" of the population, drawing heavily on folklore studies and reifying culture as nation (Cvijić 1902). The Serbian Orthodox Church revived interest in old religious buildings, especially those in Kosovo where its patriarchate was first established in the fourteenth century. The claims to an enlarged Serbian homeland weighed heavily on historical cultural sites and ideas of common descent thereby connected the expanding Serbian state to an imagined nation and territory that still lay beyond its borders.

The battlefield of Kosovo Polje, where mythologized Serb heroes resisted the Ottoman conquest, had already been popularised by the philologist Vuk Karadžić when Serbia commemorated the event in 1904 with a large monument in Kruševac, near the border of today's Kosovo. Karadžić's revisionist editing of Serbian epic ballads matched the nationalist mood, excluding Albanians and other Balkan peoples who fought with medieval Serbs against the Ottomans (Bieber 2002; Lellio and Elsie 2009). Karadžić's studies informed a later group of scholars who viewed the region's population as predominantly members of a wider Serb descent group that had adopted exotic cultural characteristics (Islam) while still proving Serbia's claim to territory (Wachtel 1998; Malcolm 1999). Thus the nationalist ideology of the expanding Serbian state at the turn of the last century was able to elide several centuries and recover an imagined medieval homeland through a complementary theory of reproductive continuity that reduced ethnic difference to "foreign" distortion of Serbian culture.

Nonetheless, Serbia's anxiety over unredeemed Serbs outside its borders mirrored its growing anxiety over the ethnic heterogeneity in its expanding national homeland. Having received international recognition as an independent state in 1878, Serbia launched a formal campaign to expel Muslims from the territories granted by the Treaty of Berlin. Violent campaigns to purge Albanians and confiscate their land pushed tens of thousands of people into Kosovo. Following Serbia's annexation of Kosovo after 1912, the government began a colonisation programme to "rebalance" its ethnic composition through a series of laws and land redistribution programmes meant to entice Serbian veterans and royalists to settle in the new province. Among the strongest advocates for colonisation and ethnic rebalancing were the head of the Serbian statistical service and nationalists of different stripes. What concerned them, in particular, was the immediate numerical disadvantage of Serbs in Kosovo and the new territories. Colonisation was seen as a rapid fix for the occupation of the new territory; natural growth would take too long (Malcolm 1999). Yet colonisation was an expensive and relatively unsuccessful programme; many Serbs emigrated away from the province, unable to make a living.

Between the world wars, impatience with the colonisation programme's slow changes to the ethnic balance led the Serbian nationalist and eminent historian Vaso Čubrilović to focus again on the differential population growth rates between Albanians and Serbs. His concern was that Albanian national-ism was better situated strategically and economically to one day reduce the size of the Serbian state. He blamed "Turkish sharia law" for the Albanians' success in Kosovo at the expense of Serbia, criticising his government's failure to use more violent means "to solve all the major ethnic problems of the troubled and bloody Balkans" (Čubrilović 2002, 98–99). Citing other coun-tries' use of population "transfers" to reduce minority peoples, he advocated the use of force in removing the "Albanian wedge" from geostrategic points deemed vital to Serbian territorial interests. Persecution and harassment of the Kosovar Albanians ensued and the campaign to remove Kosovar Alba-nians went so far as to offer payments to Turkey if they would accept a population "exchange" of Albanians, a plan stopped only by World War II. His use of terms like "cleansing" the territory to achieve purity echoed into the 1990s as the Serb nationalists employed the state apparatus and para-military forces to target non-Serb elites, forcibly evacuate key settlements, confiscate land and colonise Kosovo with Serbs.

Fertility liberalisation in socialist Yugoslavia

Whatever the Serbian designs before World War II, the occupation by Axis powers set in motion a Partisan resistance headed by Josip Broz "Tito" that led to socialist Yugoslavia (1945–1990). Serbia became one of six constituent republics in the socialist federal state in which Kosovo slowly gained greater autonomy. Like other communist parties, Yugoslavia's embraced liberation by more directly managing the means of production in opposition to the liberal and nationalist programmes of the past. As a biopolitical programme, how-ever, it was predicated on a wholly familiar modernist programme of the developmentalist state, including education, medicine, housing and material progress, all of which were gauged through regular surveys of its citizens. As a progressive movement, Yugoslav communism laid claim to improving women's lives as a necessary aim of the party and claimed a goal of gender equality, especially in education, work and political life (Ramet 2010).

In socialist Yugoslavia, fertility regulation was considered key to partici-pation in public institutions but the government took a relatively unobtrusive role in family planning, making contraceptives, induced abortions and family leave policies widely available. Gender relations within households were far more parochial, however, and contraceptive use as a means of regulating fertility lagged behind abortion in many parts of Yugoslavia (Kapor-

Stanulovic and David 1999). In effect, Yugoslav socialism achieved advances for women's reproductive rights, primarily in the public sphere, while the private sphere and traditional cultures remained quite conservative outside of major urban areas (Ramet 2010).

As the meaning of population changed through the political processes of state demography (state-istics), so too did the salience of national identity. Because the state project of Yugoslavia entailed the withering away of national identities in preparation for a common socialist future, one of its central anxieties was the persistence of what it termed "traditional", "bourgeois" and "counter-revolutionary" culture. Ironically, national identity became institutionalised in Yugoslavia because the socialist prospect of achieving a historically transcendent "New Yugoslav man" could only be measured if the enumeration of those who maintained their "traditional" identities was receding (Gruenwald 1983). Indeed, national identity's naturalness is what made it problematic, setting itself in Yugoslav socialist theory as an imaginary of historical, specifically pre-revolutionary, life (Horvat 1983). In a sense, socialist theory maintained a romanticist vision of the rustic, rooted subject of the past, who served as a necessary and convenient mythical target for the programme of a socialist future.

The socialist state's techniques of representing the ethno-territorial structure of the population directly informed the nationalists' complaints about political representation in late-Yugoslavia. We can identify these from the statistical techniques themselves. Counts of population in the statistical tables of the Yugoslav state are always followed by a subdivision of population by geographical unit and ethnicity as the primary description of distribution. The cross-tabulation of territory and identity became the foundation stones of an *episteme* that nervously measured the very ethnic identities it sought to overcome. The spatial distribution of ethnicity, along with illiteracy and other social ills, contributed to ways of seeing regions as leading or lagging in the processes of modernising political-economic change. Made visible in this way, however, ethnicity was indexed by space, becoming a social fact even as socialism endeavoured to outpace its resonance in daily life. The 1974 constitution brought greater decentralisation and the republics began to assume sovereign power over their titular nations.

Among the first policies to be "localised" were those affecting fertility as the more Catholic republics of Slovenia and Croatia restricted access to contraceptives and abortions, in keeping with Vatican teaching. Sex education, meanwhile, was taught in theory but not in the classroom, especially in rural areas. Family planning techniques were encouraged by the Serbian government, especially in Kosovo among Muslims (Kapor-Stanulovic and David 1999). Serbs in Kosovo, however, complained of the higher birthrates among the larger Albanian population, which was predominantly Muslim but also Catholic. Community leaders feared that the growing autonomy granted to the province after the 1960s meant that land, jobs and resources were being taken away from Kosovar Serbs. It was in addressing these points that Milošević found his nationalist voice in 1987, telling the crowd gathered not far from the mythic battlefield of 1389, "nobody will beat you".

Not only were ethnic categories necessarily categories of political thought in Serbia and Yugoslavia, their numericisation and spatialisation through censuses and maps enhanced their politicisation (Rose 1991; Legg 2005). The relative availability of Yugoslav statistics was unusual in an authoritarian socialist state, and these numbers became "mobile and polyvocal resources" for the critics of state socialism, especially nationalists (Rose 1991, 684). Curiously, given the attention to nineteenth and twentieth century Serb nationalist justifications for a state based on a numerically superior Serb population, Serb nationalists in the 1980s and 1990s were focused on the statistical and political inferiority of their position in Serb lands. Statistical computations of population composition and fertility characteristics added to the medical data on the low birthrate in many republics, except among Muslim communities. Yugoslavia's state cartographers produced numerous maps of ethnic distribution that directly informed the belligerents who sought to draw new national boundaries among the peoples of Yugoslavia (Campbell 1999). The calculation of ethnically differentiated birthrates allowed nationalists to project a future national dystopia for the "Serb lands". The fear of becoming a minority nation in both Yugoslavia and Serbia itself was a mobilising anxiety for Serb nationalists, among them Milošević's wife and other Belgrade elites.

The ultimate dissolution of Yugoslavia beginning in 1991 could thus be cast as a culmination of nationalist biopolitical rejection of the federal socialist state's developmentalist programme to improve the conditions of life for national populations. Popular confrontations and protests against the socialist regime came to a head in competitive elections that raised what Foucault's later works describe as liberalism's question "Is one governing too much?", but there was no liberal answer. The corruption and economic mismanagement in late Yugoslav socialism was proof for some that the socialist system was too cumbersome and inefficient. Redistribution policies between federal units fueled resentment in the larger republics and initiated a quest for further autonomy, even independence. These complaints implied a more pressing question, "Who governs whom?" For most republics, the simple fact of their large national majorities and the electorates' demand in 1990 for economic reform spawned nationalist parties focused on economic liberalisation and national unification. In this context, biopolitical thinking begot ethnomajoritarian demands.

Biopolitical competition and war

Late socialist Yugoslavia witnessed a reckoning over the unresolved questions of territory inhabited by a multi-national population. It would be a mistake, however, to view the nationalist movements that tore apart Yugoslavia as continuities of their nineteenth-century forerunners. Neither were these movements wholly new but instead trafficked in the modern state's institutional narratives of territory and nation, which had first appeared at a time when the geopolitical order was structured around wholly different epistemologies of territoriality and population. What the violence of the 1990s represented was a new historiographical and geopolitical telling of how identity, belonging and space could yet be reconfigured to achieve a national future in independent states, entities that had never before existed in that region. This new narrative, that every nation should have its own state, was largely alien to Yugoslavs on the eve of the war just as it had been during the many centuries in which multi-confessional and polyglot communities defined a modus vivendi throughout Eastern Europe (Hodson et al. 1994; Mahmut-ćehajić 2000).

Central to the geopolitics that emerged in the 1980s and 1990s was the role that reproductive competition, or reproductive geopolitics, played in the debates and the war plans that led to the wars in Croatia, Bosnia and Herzegovina, Kosovo and Macedonia. Contemporary politicians were laying claim to national territories while, at the same time, denying the stability of the boundaries that had been established at that time. Indeed, the argument that the existing boundaries of the Yugoslav republics were artificial political constructs became a prerequisite for credibility in the nationalist movements. While the European Community's arbitration of Yugoslavia's dissolution argued for the durability of independent republican borders, the plans for war explicitly called for territorial alterations on the basis of perceived historical or contemporary majorities. Political claims to revising national territory and endless rebuttals appeared in major newspapers along with maps that showed border changes and population counts, computational representations impossible until the late 1980s. The technological infusion of biopolitical data into the geopolitical imagination of nation created more than just debate via maps.

More profoundly, the mobile and polyvocal data on ethnic structure and fertility were generative of three wider biopolitical questions that fueled the geopolitical imagination of national futures. First came the question of minoritisation, the arrival of a future in which one's people are rendered statistically and politically obsolescent. The programme of "national balance" earlier in the twentieth century had responded primarily to the idea of occupation and productive use of the land through colonisation and expulsion. As socialist Yugoslavia's statistical offices generated more complete calculations of fertility and growth projections, a different anxiety took hold. Productive use of the land became relatively less important than employment rates and income levels as nineteenth-century physiocracy gave way to urban industrial development. For nationalists, statistical irrelevance meant dissipation and exhaustion, amalgamation and assimilation. Reproduction, on the other hand, promised new national subjects whose biopower could secure the nation's future. Serbian nationalism, therefore, was concerned with inculcating biopolitical anxiety as a mobilisation of national resources that could restore the nation amid the failing institutions of late Yugoslavia.

The appearance of competitive elections in Eastern Europe in the late 1980s and Yugoslavia in 1990 lent urgency to a second biopolitical question of how to organise a political party that could achieve democratic victory in a highly decentralised state. Crosscutting parties such as socialists, social democrats and liberal democrats were seen as too close to the corrupt and discredited state officials of Yugoslav communism. The assumption of titular sovereignty in the republics thus fueled both legal arguments and nationalist politics towards independence. Republic-level parties appealed to the titular majorities that dominated most of the republics and provinces. Slovene, Croat and Serb "democratic" parties defeated the ruling communist parties in the republics, except in Serbia where Slobodan Milošević led the Socialist Party to victory by arguing that the other republics' desire for independence threatened to divide the Serbian people. Only Bosnia and Herzegovina lacked a clear majority but the Croatian and Serbian "democratic nationalist" parties were very popular among their respective communities. Economic and social instability drove support for independence, particularly after Milošević attempted to centralise power over the other republics. Whereas territory was once national destiny, it was now national identity that was political destiny—territory could be altered to match.

The fear of minoritisation through competitive reproduction and the politicisation of national electoral power gave rise to a third biopolitical question of how violence would be used to end the political stalemate in Yugoslavia. The wars that accompanied the dissolution of Yugoslavia have been branded "new wars" because they targeted not territory or security apparatuses, per se, but the population itself (Kaldor 1999). The wars in Croatia, Bosnia and Herzegovina, Macedonia and Kosovo were each part of wider nationalist campaigns to permanently alter the demographic landscape and establish new republic borders that would preserve national majorities. The pernicious logic behind the war, that of national survival and majoritarianism, was evident in the Serbian plans to carve out territory from the newly independent Croatia and Bosnia and Herzegovina based largely on demographic claims to land and statistical superiority, either contemporary or historical (Jansen 2005). The Serb forces intentionally murdered and expelled citizens on the basis of their non-Serb identity, accompanied by the destruction of their cultural sites and the importation of colonising Serbs. These "ethnic cleansing" operations in Bosnia and Kosovo were not unlike what Čubrilović had prescribed in 1937, a war on the lives of others, per se. This is why the wars in Yugoslavia were genocidal in intent (Dahlman 2005).

Importantly, the biopolitical *episteme* that made fertility and national growth projections central to the nationalists' vision of their futures consequently targeted women's bodies as sites of a programme to directly manage ungovernable bodies. Control of Serbian women and their reproductive capacity was indeed the point of the term "white plague" made popular by the Serbian Orthodox Church in its 1994 Christmas message, which described three sins by Serbian women that were destroying the nation: the sin of having no or only one child, the sin of becoming a minority and the sin of abortion. Directed at other groups, the same logic made systematic use of rape warfare, especially by Serb forces. Rape in this context was not a secondary effect of war but an organised tactic, a fulfillment of the biopolitical imperative to "abolish old forms of life and create new ones" (Prozorov 2013, 211). The intended purpose of the rape camps included the sexualisation of punishment and sexual trafficking, as well as a policy of forced impregnation of Muslim and Catholic women (Salzman 1998). The policy of enforcing these births illustrates the nationalist's biopolitical reasoning that forcing Serbian bloodlines into the victim's communities would destabilise the next generation's sense of their national identity (Benderly 1997; Weitsman 2008).

Conclusion

This chapter has provided a specific history of how biopolitics figured in the development of Serbian nationalism since the nineteenth century. By focusing on the means by which the modern Serbian state has come to make its population knowable, legible and calculable, we may begin to catch a glimpse of how the formation of a biopolitical *episteme* renders the problematic of population growth, fertility and ethnicity as specific expressions of territorial insecurity amid national competition. Returning to the earlier discussion of the "white plague", it comes as no surprise that the term appeared not in the 1950s, when Serbian birthrates began to decline, but in the biopolitically legible context of fertility rates, ethnic majorities and reproductive competition in late Yugoslavia (cf. Jansen and Helms 2009).

National survival, the security of Serbia's future as a people, had thus become a wholly ideological construct of gender relations that limited women to the role of naturalistic progenitors of the nation, rather than subjects of social and political rights and agents of transformative change who made their own reproductive choices. What remains striking in this history is not that political ideologies adopted biopolitical ways of knowing and intervening in state populations. Instead, it is remarkable that biopolitics emerged in Serbia more or less in tandem with Western liberal states despite the radical ideological shifts in Serbian politics over the last two centuries. Yet what would be more remarkable still, and perhaps even liberating, would be a politics of human life that abandoned the capture and discipline of biopower as the goal of the state.

Notes

1 The Women in Black were most active in late Yugoslavia, protesting against war in the 1990s by highlighting its effects on women in Serbia and the other republics.

2 Serbia's total fertility rate was over 3.5 births per woman after World War II and declined to below replacement level (2.1) during the 2000s. For the ethnically Serb population, relative decline was evident from the 1960s and absolute decline was statistically apparent in the 1980s.
3 The Turks living in the region were not even enumerated, treated as ephemera of the receding Ottoman Empire and under orders to evacuate the principality.

Acknowledgements

The author would like to thank the editors, Emily Channell-Justice, and Elana Resnick for their insightful and constructive suggestions.

References

Agamben, Giorgio. 1998. *Homo Sacer*. Stanford, Calif: Stanford University Press.
Anderson, Benedict. 2006. *Imagined Communities: Reflections on the Origin and Spread of Nationalism*. Verso.
B92. 2018. "Deputy PM Denounces Ministry's Slogans as Offensive to Women," February 14, 2018.
Bashford, Alison. 2016. *Global Population: History, Geopolitics, and Life on Earth*. New York: Columbia University Press.
Benderly, Jill. 1997. "Rape, Feminism, and Nationalism in the War in Yugoslav Successor States." In *Feminist Nationalism*, edited by Lois A. West, 57–72. New York: Routledge.
Bieber, Florian. 2002. "Nationalist Mobilization and Stories of Serb Suffering: The Kosovo Myth from 600th Anniversary to the Present." *Rethinking History* 6(*1*): 95–110.
Campbell, David. 1999. "Apartheid Cartography: The Political Anthropology and Spatial Effects of International Diplomacy in Bosnia." *Political Geography* 18(*4*): 395–435.
Čubrilović, Vaso. 2002. "The Expulsions of the Albanians Memorandum." In *Gathering Coulds: The Roots of Ethnic Cleansing in Kosovo and Macedonia*, edited by Robert Elsie, 97–130. Pejë: Dukagjini Publishing House.
Cvijetić, Leposava. 1984. "Popis Stanovništva i Imovine u Srbiji 1834. Godine." *Miscellanea 13*: 9–118.
Cvijić, Jovan. 1902. "Antropogeografski Problemi Balkanskog Poluostrva." Serbian Academy of Sciences: Belgrade.
Dahlman, Carl T. 2005. "Geographies of Genocide and Ethnic Cleansing: The Lessons of Bosnia-Herzegovina." In *The Geography of War and Peace: From Death Camps to Diplomats*, edited by Colin Flint. New York: Oxford University Press.
Deutscher, Penelope. 2008. "The Inversion of Exceptionality: Foucault, Agamben, and 'Reproductive Rights.'" *South Atlantic Quarterly* 107(*1*): 55–70.
Dnevnik. 2018. "Љубав и беба-првё штё нам треба! Рађај, не ёдгаћај!" February 13, 2018.
Dragović-Soso, Jasna. 2004. "Rethinking Yugoslavia: Serbian Intellectuals and the 'National Question' in Historical Perspective." *Contemporary European History* 13 (*2*): 170–184.
Elden, Stuart. 2007. "Governmentality, Calculation, Territory." *Environment and Planning D: Society and Space* 25(*3*): 562–580.
FBIS Daily Report 27 April 1987. "Milosevic Address," April 25, 1987.
Foucault, Michel. 2003. *"Society Must Be Defended": Lectures at the Collège de France, 1975–76*. New York: Picador.

Foucault, Michel. 2007. *Security, Territory, Population Lectures at the Collège de France, 1977–78.* New York: Palgrave Macmillan.

Gruenwald, Oskar. 1983. *The Yugoslav Search for Man: Marxist Humanism in Contemporary Yugoslavia.* South Hadley, Mass.: J.F. Bergin.

Halpern, Joel Martin and Barbara Kerewsky-Halpern. 1972. *A Serbian Village in Historical Perspective.* Prospect Heights, Ill.: Waveland Press.

Hehn, Paul N. 1975. "The Origins of Modern Pan-Serbism: The 1844 Nacertanije of Ilija Garasanin." *East European Quarterly 9(2)*: 153–171.

Hodson, Randy, Dusko Sekulic and Garth Massey. 1994. "National Tolerance in the Former Yugoslavia." *American Journal of Sociology 99(6)*: 1534–1558.

Horvat, Branko. 1983. *The Political Economy of Socialism: A Marxist Social Theory.* London: Sharpe.

Jansen, Stef. 2005. "National Numbers in Context: Maps and Stats in Representations of the Post-Yugoslav Wars." *Identities 12(1)*: 45–68.

Jansen, Stef and Elissa Helms. 2009. "The 'White Plague': National-Demographic Rhetoric and Its Gendered Resonance after the Post-Yugoslav Wars." In *Gender Dynamics and Post-Conflict Reconstruction*, edited by Christine Eifler and Ruth Seifert, 219–243. Frankfurt am Main: Peter Lang.

Kaldor, Mary. 1999. *New and Old Wars: Organized Violence in a Global Era.* Cambridge, U.K.: Polity Press.

Kapor-Stanulovic, Nila and Henry David. 1999. "Former Yugoslavia and Successor States." In *From Abortion to Contraception: A Resource to Public Policies in Central and Eastern Europe from 1917 to the Present*, edited by Henry David and Johanna Skilogianis, 279–315. Westport, Conn.: Greenwood Press.

Kligman, Gail. 1998. *The Politics of Duplicity: Controlling Reproduction in Ceausescu's Romania.* Berkeley: University of California Press.

Legg, Stephen. 2005. "Foucault's Population Geographies: Classifications, Biopolitics and Governmental Spaces." *Population, Space and Place 11(3)*: 137–156.

Lellio, Anna Di and Robert Elsie. 2009. *The Battle of Kosovo 1389: An Albanian Epic.* New York: I.B. Tauris.

Lemke, Thomas. 2001. "'The Birth of Bio-Politics': Michel Foucault's Lecture at the Collège de France on Neo-Liberal Governmentality." *Economy and Society 30(2)*: 190–207.

Lemke, Thomas. 2011. *Biopolitics: An Advanced Introduction.* New York: New York University Press.

Mahmutćehajić, Rusmir. 2000. *Bosnia the Good: Tolerance and Tradition.* Budapest: Central European University Press.

Malcolm, Noel. 1999. *Kosovo: A Short History.* New York: HarperPerennial.

Mostov, Julie. 2000. "Sexing the Nation/Desexing the Body: Politics of National Identity in the Former Yugoslavia." In *Gender Ironies of Nationalism: Sexing the Nation*, edited by Tamar Mayer, 103–126. New York: Routledge.

Palairet, Michael. 1995. "Rural Serbia in the Light of the Census of 1863." *Journal of European Economic History 24(1)*: 41–107.

Papić, Žarana. 2010. "Women in Serbia: Post-Communism, War, and Nationalist Mutations." In *Gender Politics in the Western Balkans: Women and Society in Yugoslavia and the Yugoslav Successor States*, edited by Sabrina P. Ramet, 153–169. University Park: The Pennsylvania State University Press.

Petryna, Adriana. 2013. *Life Exposed: Biological Citizens After Chernobyl.* Princeton, N.J.: Princeton University Press.

Porter, Theodore M. 1986. *The Rise of Statistical Thinking, 1820–1900*. Princeton, N. J.: Princeton University Press.

Prozorov, Sergei. 2013. "Living Ideas and Dead Bodies: The Biopolitics of Stalinism." *Alternatives: Global, Local, Political 38(3)*: 208–227.

Ramet, Sabrina P. 2010. "In Tito's Time." In *Gender Politics in the Western Balkans: Women and Society in Yugoslavia and the Yugoslav Successor States*, edited by Sabrina P. Ramet, 89–105. University Park: The Pennsylvania State University Press.

Rancière, Jacques. 2004. *The Politics Of Aesthetics: The Distribution of the Sensible*. London: Continuum.

Richardson, Diane. 2017. "Rethinking Sexual Citizenship." *Sociology 51(2)*: 208–224.

Richardson, Eileen and Bryan S. Turner. 2001. "Sexual, Intimate or Reproductive Citizenship?" *Citizenship Studies 5(3)*: 329–338.

Rose, Nikolas. 1991. "Governing by Numbers: Figuring out Democracy." *Accounting, Organizations and Society 16(7)*: 673–692.

Rose, Nikolas. 2007. *Politics of Life Itself: Biomedicine, Power, and Subjectivity in the Twenty-First Century*. Princeton: Princeton University Press.

Rose, Nikolas and Peter Miller. 2010. "Political Power beyond the State: Problematics of Government." *British Journal of Sociology 61(s1)*: 271–303.

Ryder, Norman B., Philip M. Hauser and Wilson H. Grabill. 1971. "Notes on Fertility Measurement." *The Milbank Memorial Fund Quarterly 49(4)*: 109–131.

Salzman, Todd A. 1998. "Rape Camps as a Means of Ethnic Cleansing: Religious, Cultural, and Ethical Responses to Rape Victims in the Former Yugoslavia." *Human Rights Quarterly 20(2)*: 348–378.

Savić, Bojan. 2014. "Where Is Serbia? Traditions of Spatial Identity and State Positioning in Serbian Geopolitical Culture." *Geopolitics 19(3)*: 684–718.

Shaw, Stanford J. 1978. "The Ottoman Census System and Population, 1831–1914." *International Journal of Middle East Studies 9(3)*: 325–338.

Stokes, Gale. 1990. *Politics as Development: The Emergence of Political Parties in Nineteenth Century Serbia*. Durham, N.C.: Duke University Press.

Turner, Bryan S. 2003. "The Erosion of Citizenship." *The British Journal of Sociology 52(2)*: 189–209.

Turner, Bryan S. 2008. "Citizenship, Reproduction and the State: International Marriage and Human Rights." *Citizenship Studies 12(1)*: 45–54.

Vuletić, Aleksandra. 2012. "Censuses in 19th Century Serbia: Inventory of Preserved Microdata." 2012–2018. MPIDR Working Paper. Rostock: Max Planck Institute for Demographic Research.

Wachtel, Andrew. 1998. *Making a Nation, Breaking a Nation: Literature and Cultural Politics in Yugoslavia*. Stanford, Calif.: Stanford University Press.

Weitsman, Patricia A. 2008. "The Politics of Identity and Sexual Violence: A Review of Bosnia and Rwanda." *Human Rights Quarterly 30(3)*: 561–578.

Zajovic, Stasa. 1995. "Birth, Nationalism and War." available at http://www.hartford-hwp.com/archives/62/039.html, last accessed 8 Aug. 2018.

Žarkov, Dubravka. 2007. *The Body of War: Media, Ethnicity, and Gender in the Break-up of Yugoslavia*. Durham, N.C.: Duke University Press.

Zarkovic, S. S. 1955. "Sampling Methods in the Yugoslav 1953 Census of Population." *Journal of the American Statistical Association 50(271)*: 720–737.

11 Reproducing inequalities

Examining the intersection of environment and global maternal and child health

Andrea Rishworth and Jenna Dixon

Situated in the discipline of health geography, this chapter examines maternal and child health issues at the interface of environmental health. Underpinned by the premise that one's health and geographies are inextricably linked, this work draws from a case study in Ghana's rural north to examine how precarious environments underscored by food insecurity, fluctuating climates, historically rooted inequality and poverty contribute towards dubious maternal and child health outcomes. This work comes in part as a response to the global failures in addressing maternal and child health problems, which have led some to declare that global health is "tipping into irrelevance". While global maternal and child health movements have tended to produce programmes blind to these interactions, we draw from health geography and perspectives of political ecology to argue for a reawakening in health policy and research praxis that engages equally with environmental issues, and indeed recognises that maternal and child health cannot be realised without addressing the health of place.

Global maternal, newborn and child health agendas

Matters of maternal, newborn and child health (MNCH) have for many years played prominently in global health agendas. From the World Summit for Children (1990), the Safe Motherhood Initiative (1987), the International Conference on Population and Development (1994) and Millennium Development Goals 4 and 5 (MDGs), the interconnected health needs of women and children have been widely recognised. In theory, these global agendas lay the groundwork for comprehensive MNCH approaches, yet in practice, complicated ideological and programmatic differences between competing MNCH interest groups have led to polarised—and largely ineffective—global MNCH agendas (Starrs 2014; Kabeer 2015).

Positioning maternal and child health as equally important in the MDGs enabled the MNCH groups to form "The Partnership for Maternal, Newborn & Child Health" (Storeng and Béhague 2014). While the rhetoric of a unified MNCH approach was intended to enhance health outcomes through a seamless delivery of health interventions over time (from pre-pregnancy to pregnancy, delivery and early childhood), and space (from facility, community, regional, national policy scale) (PMNCH 2009), the medicalised thrust of the MDGs (see Fehling et al. 2013) positioned maternal-neonatal and child mortality as the dominant "problem", in which the prescription was more money, better policy and programme support (see McDougall 2016). As a result, health systems were framed as technical delivery channels, providing commodities, interventions and workers, rather than as social institutions shaped for and by the interaction of people, policies and services.

The transition to the Sustainable Development Goals (SDGs) marks a new chapter in which to reconceptualise issues of MNCH. Born post-MDGs, the 17 SDGs set by the United Nations guide policy and funding for the next 15 years, by focussing on the interconnected social and economic challenges facing the planet. For our discussion, we note Goal 3 which aims to "ensure healthy lives and promote wellbeing for all at all ages" (UNSEA 2015), providing the opportunity to rethink how health is shaped across the life course and intergenerationally. By placing greater emphasis on the multidimensional nature of health, the SDGs offer a way to move beyond numeric reductions in maternal mortality or access to reproductive healthcare, and instead consider how maternal health is shaped through the interaction of biological, social and environmental factors.

Despite this opportunity, overwhelming focus remains on biomedical elements of mortality reduction (Graham et al. 2016), precluding consideration towards the broader range of inputs and outcomes for MNCH and its implications for intervention (Langer et al. 2015). While the MNCH community certainly understands the biomedical characteristics of unhealthy pregnancy (e.g., anaemia, hemorrhage), little consideration is given to the influence of mediating factors (e.g., knowledge, family structure), broader determinants (e.g., wealth, distance, transport) and the availability of secondary or tertiary prevention (Graham et al. 2016).

Indeed, "a vast unfinished agenda in maternal and reproductive health" remains (Kassebaum et al. 2017, 697). Globally, while maternal mortality rates have declined, global maternal death rates for the poorest are on the rise (Kassebaum et al. 2016). Concurrently, geographic disparities between and within high and low-income countries for maternal, infant and child mortality have widened (see Kassebaum et al. 2016), with the highest death rates in sub-Saharan Africa. The inability to quell such persistent inequities has led to the broader assertion that global health is "tipping into irrelevance", as Horton (2016, 1362) states,

> Caroline Maposhere, a nurse midwife from Zimbabwe, won applause when she pointed out that, "A woman doesn't live her life in compartments". She described how pregnant women are forced to give birth on the floor, provide their own water, and wash their own laundry. But there were few signs that leaders of the women's and children's health movement recognised the challenge Maposhere was making. The global health

community has become lazy, self-regarding, and conceited. It is no longer shocked by the injustices it sees. As it talks among itself, it is tipping into fatal irrelevance.

Our goal in this chapter is to suggest that approaches and conceptual frameworks of (health) geography are needed to inform the formulation of new MNCH priorities in the SDG era. Using a case study from Ghana's Upper West Region (UWR) we show how and why new iterations and approaches to MNCH must consider both social and ecological factors in order to truly enhance opportunities and outcomes of maternal health and wellbeing. We further seek to understand the embodied experiences and long-term impacts of pregnancy for women and children. Empirical evidence from two case studies examining issues of gender, health and health insurance schemes in rural UWR are employed. While the aims of our two studies focused on gendered health experiences in relation to different aspects of health insurance in the UWR, we both inductively found the need to consider issues of health in place, understanding how social, political and ecological factors shape maternal health and wellbeing (see Rishworth 2014; Dixon 2014). In-depth field work with female residents provided a unique opportunity to consider the following questions: (1) what factors other than primary healthcare contributes to MNCH; (2) what are the gendered and embodied experiences of the environment that contribute to MNCH?

Conceptually, we answer these questions by employing human-environmental research in political ecology of health (King and Crews 2013; Jackson and Neely 2015; King 2015, 2017), integrated with feminist political ecology (Rocheleau et al. 1996; Sultana 2012; Elmhirst 2015). We use the case study of maternal health to demonstrate and advance knowledge on two interrelated fronts. First, the case study shows how and why political ecology's multiscale analysis of socioeconomic and political processes of human environmental relations must be extended to consider the role of gendered bodies in the production, experiences and outcomes of health, recently advanced by geographers (see Guthman 2012; Sultana 2012; Guthman and Mansfield 2015; King 2017). Second, this case study demonstrates how these perspectives would immeasurably strengthen new MNCH policies and approaches in the SDG era.

Theoretical perspectives of health and place

Maternal, newborn, child and indeed all human health, is expressed, experienced and exchanged at the nexus between social and environmental systems (King and Crews 2013). To understand this complex nexus, health geography and approaches of political ecology provide important conduits to elucidate the ways in which health is embedded in the biophysical world (Richmond et al. 2005; Jackson and Neely 2015). Given that health is shaped by a combination of sociocultural, political-economic and biological factors that vary across space and time (King and Crews 2013; Gatrell and Elliott 2014), "health"—a state of total mental and physical wellbeing—is always context-specific, shaped and embedded within larger environment systems (King 2015).

Political ecology of health

Theoretically rooted in the interdisciplinary lens of political ecology (Robbins 2011), political ecology of health (PEH) has played a pivotal role in bringing attention to the ways health is shaped though the environments in which people live, as well as how human-nonhuman interactions at different geographical scales influence health (Richmond et al. 2005; King 2010, 2015; King and Crews 2013). As an analytic framework, PEH examines human health in the context of broader social and economic dynamics, paying particular attention to historical dynamics and the role of power (e.g., colonial history, development policies, global guidelines) (King 2010). PEH is also inherently complementary with health geography, as the theory outlines that health can only be understood with explicit consideration towards the environment. Indeed, it is only at "the nexus between social and ecological processes that human health is shaped, and only in understanding the interaction between them can human health be properly understood" (King 2015, 349). The explanatory power of PEH straddles three interrelated themes: (1) the examination of health and wellbeing within broader socioecological contexts; (2) multi-scale analysis that engages scales of the individual human body to the macro-scale structural forces; (3) sensitivity to the historical evolution and dynamics of contemporary health outcomes in a given place (King and Crews 2013; King 2015; Neely 2015).

PEH is critical of biomedical approaches and instead advocates for a more integrated perspective on human health and wellbeing, embedded within socioecological and political systems (see Richmond et al. 2005; King 2010). Transitioning from structural approaches (e.g., Richmond et al. 2005) to a more variegated post-structural consideration of biophysical environmental relations and health (e.g., Neely 2015; King 2017), PEH literature has substantially improved knowledge on the pertinent ways macro processes intersect with human populations to produce (ill)health and (ill)wellbeing.

Notwithstanding, PEH has some limitations. While multi-scale analysis is critical to PEH, little work unpacks the finer scales of the individual human body (Guthman and Mansfield 2013). This is emblematic of political ecology more broadly, retaining focus on community landscapes, rather than individual bodies (Guthman and Mansfield 2013; King 2015). Though some new geographic offshoots are beginning to examine issues of the body (e.g., feelings and affect [Hayes-Conroy and Hayes-Conroy 2015], obesity and biological changes [Guthman 2012]), recent reviews have called for more work on the critical political ecologies of health and the body (Guthman and Mansfield 2013; King 2015).

Moreover, with few notable exceptions (e.g., Sultana 2012), much existing PEH scholarship remains ungendered. While gender is a key theme in political ecology writ large (Elmhirst 2015), gendered analysis in PEH is fairly limited (Jackson and Neely 2015; King 2015, 2017). While successfully eluci-dating how macro-structural forces rework ecological processes in ways that influence new health outcomes (e.g., King and Crews 2013), PEH remains silent on how resulting health outcomes are gendered, not to mention how gendered bodies intersect within inter-intra household and community rela-tions. Although it is important to understand how social ecology and politics shape health at broader scales (e.g., neighbourhoods, municipalities and nation-states), it is equally imperative for PEH scholars to expand considera-tion of politics into household spheres and the gendered bodies therein.

Towards a feminist political ecology

Feminist political ecology (FPE) emphasises how gendered power relations are informed by politics and power at differential scales (Rocheleau et al. 1996). Extending the scope and scale of political ecology, FPE examines the "everyday" embodied experiences in local environments within the context of global processes of environmental and economic change (Rocheleau et al. 1996; Elmhirst 2015). Accordingly, FPE "treats gender as a critical variable in shaping resource access and control, interacting with class, caste, race, cul-ture, and ethnicity, age, et cetera, to shape processes of ecological change" (Rocheleau et al. 1996, 4). As a result, FPE attends to the ways multiple social differences are reproduced through everyday embodied practices, avoiding simplistic gendered divisions in human environment relations (Elm-hirst 2015). Further, FPE's critical analysis of scale in political ecology offers a disaggregated and gendered account of "what has been called 'community' and 'local', and the often-homogenous unit of the household" (Rocheleau 2008, 722). In doing so, no scale is privileged over the other; instead, scholars recognise that fluid sets of relations between individuals and households intersect with broad scale dynamics to shape human-environmental relations.

"Though emerging interests within natural and social science center on the social and ecological dimensions of human health (and disease), there have been few studies that address them in an integrated manner" (King and Crews 2013, 1). Accordingly, this chapter incorporates PEH and FPE approaches to examine social and ecological dimensions of maternal health. In doing so we put forth a feminist political ecology of health (FPEH) approach to question how framings of MNCH in community, national and global politics shape opportunities and outcomes of maternal health and wellbeing. Equally, a FPEH approach questions the relations between social and environmental systems to elucidate the everyday embodied lives of women and children. Although one SDG focuses on health, many other goals—environment, nutrition, hunger, sustainable production and consump-tion, agriculture and education—directly or indirectly relate to MNCH, necessitating a systems thinking approach to health. Thus, this chapter seeks to address the political ecology of MNCH and its implications in the context of the SDGs through a case study on Ghana's Upper West Region.

Study context: The Upper West Region of Ghana

The Upper West Region (UWR) spans 18,476 km^2, 7.7 percent of the total land area in Ghana. With 702,110 citizens, the UWR has a lower population density than other regions in the country (GSS 2013). The region is one of the poorest in the country, with poverty rates two to three times higher than the national average. Explanations for such inequalities are rooted in geographically uneven policies of the colonial state that exploited the region as a labour reserve and broadly underfunded social services such as healthcare, a trend that has been perpetuated throughout post-colonial regimes (Songsore 2003).

Located within the semi-arid Guinea Savannah belt, the UWR experiences one rainy season from late May to early September, followed by seven to eight months of dry weather. In comparison, the more fertile regions in Ghana's south benefit from multiple rainy seasons and the ability to produce cash crops such as cocoa. Over 80 per cent of the population in the UWR is engaged in agriculture (compared to the national average of 51 per cent [GSS 2013]), yet regional climatic and agro-ecological factors leave 63.6 percent of the population severely food insecure (Atuoye and Luginaah 2017). Prevalence of anaemia is 98.7 per cent among infants (1–2 years), probably associated with high rates of maternal micronutrient deficiency (Ewusie et al. 2014). Malnutrition is almost twice the national average (Glover-Amengor et al. 2016).

While the family forms the bedrock of Ghanaian society (Takyi and Oheneba-Sakyi 2006), within the family household, resources and responsibilities are commonly split along gendered lines, which means the household system cannot be regarded as a simple unified economy (Agarwal 1997). Men, as household heads in the patriarchal UWR, are expected to provide financial support for the family and are given more societal leeway for personal spending and leisure, whereas women are often expected to take their earned income and free time and put it towards their motherly responsibilities including health, food and care provision (Tolhurst et al. 2008). In most communities in the region, including those where the study was conducted, the kinship system is patrilineal, and the organisation of kinship structure around patrilines often marginalises women (Ganle et al. 2016). On top of this, growing economic and livelihood stress have exacerbated women's pre-existing responsibilities and gendered roles (Tsikata and Awetori Yaro 2014; Nyantakyi-Frimpong and Kerr 2017). As argued by a number of researchers, "gender-based role differentiation and male domination are pervasive in the UWR" (Ganle et al. 2016, 197).

In 2003, Ghana introduced its National Health Insurance Scheme (NHIS)—a district level social health insurance scheme. In principle, the NHIS offers a nationally recognised, heavily subsidised mechanism for citizens to obtain health coverage without the risk of catastrophic health spending. The scheme was built to be "pro poor" with many vulnerable groups, such as the elderly, children, social security contributors and pensioners, exempt from paying premiums (GNHIA 2013). Yet until 2008, pregnant women were still required to pay an annual premium. In part motivated by the global political agenda to address MNCH, reduce maternal mortality and enhance gender equity, Ghana implemented a Maternal Exemption Policy (MEP) under its NHIS, effectively granting free health services for all pregnant women for a one-year period (Dzakpasu et al. 2012). The MEP also affords the newborn child healthcare services for three months after birth. However, at the end of the three-month postnatal period, women previously not enrolled in national health insurance are no longer able to receive free health services and additional steps must be taken for the child to qualify for premium exempt enrolment.

Since the implementation of the MEP, Ghana has witnessed notable declines in maternal mortality and under five mortality rates (WHO 2012). Yet, maternal and child health inequalities persist and continue to grow (ibid.). While the program allows all women to enrol and use maternal health services, large equity gaps remain in terms of health facility delivery, access to skilled birth attendants, postnatal care, cesarean section delivery and access to interventions along the continuum of care. Evidence shows that it is the poor, least educated and rural women who are predominantly excluded from this care (GSS 2012). The burden of long and expensive journey times to health facilities, gendered care obligations and growing environmental concerns (such as food shortages, drought, etc.) further complicate maternal and child health outcomes (ibid.).

Methods

The data for this chapter come from two separate research projects, conducted in 2011 and 2013, focused on gendered access and utilisation of the National Health Insurance Scheme in Ghana's UWR. Both studies *inductively* found the need to consider issues of health in place, understanding how social, political and ecological factors shape maternal health and wellbeing. As such, the results below are a re-analysis of the data from the two studies with a specific focus on: (1) factors other than primary healthcare which contribute to MNCH, and (2) the gendered and embodied experiences of the environment that contribute to MNCH. In keeping with the methodological tradition of political ecology, both studies employed an intensive qualitative research approach, emphasising the local groundedness and context balanced with the recognition of larger structural factors (Robbins 2011; Miles et al. 2014). Drawing from feminist political ecology brought to the fore additional dimensions of social relations and complexities of daily life (e.g., intra-household, intra-community gender politics) in which to examine women's maternal health (see Rocheleau et al. 1996).

Results from in-depth interviews and focus groups are presented from both studies and should be interpreted as a whole. To examine women's perceptions and experiences of maternal health vis-à-vis the environment, we adopted an interpretive approach conducted across the geographical space of the UWR. The focus group serves to depict processes underlying successful participant education, as group members can ask questions, exchange anecdotes and comment on others' experiences and points of view (Kidd and Parshall 2000). Focus groups are particularly appropriate for facilitating discussions of unfamiliar topics because the less inhibited members of the group often break the ice for shyer participants (ibid.). In-depth interviews complement this method by allowing quieter participants to speak and assure anonymity (Hessey-Biber 2003).

The NHIS study (2011) included focus groups (n = 17) and in-depth interviews (n = 26) with female and male participants over the age of 18 (total = 211 participants). The MEP study (2013) included focus groups (n = 10) and in-depth interviews (n = 30) with females 18 years and older (total = 147 participants). These provided a range of experiences and perceptions of health in various locations. All in-depth interviews were conducted in privacy and the participants were assured anonymity. In-depth interviews varied in length between one half hour and three hours. For both studies, all focus groups were conducted outdoors under trees, with participants grouped into either female or male groups to provide social comfort; they averaged approximately one to one and a half hours in length.

In-depth interviews and focus groups with women aged 18 and over were digitally recorded, with the consent of the respondents. The in-depth interviews and focus groups were led by the researcher and translated by a research assistant fluent in Dagaare or Waali (the languages of the participants). A checklist of topics (semi-structured open-ended questions) was used to probe participants about several different topics related to their experiences with and opinions about health, maternal health concerns, the MEP and the NHIS in the UWR.

Analysis

Despite the primary focus of both studies being access to healthcare, many participants spoke of the importance of the broader physical, sociocultural and political environment on MNCH. As such, for the purposes of this chapter, we performed an additional analysis to highlight and further explore these inductively important themes. The two authors on this chapter conducted this analysis and were also the investigators of the original studies, thus ensuring qualitative rigour. We began by both independently reviewing four random transcripts from each study (eight total) in order to confirm there were factors in the data which related to MNCH other than primary healthcare. Once confirmed, the authors worked together on a full thematic analysis of the data.

The thematic analysis involved inductive and deductive coding of the data. First, we performed a detailed review of the randomly selected transcripts to identify initial themes that reflected important aspects of MNCH other than primary healthcare and specifically linked to the gendered and embodied experiences of the environment that contribute to MNCH. Second, we drafted a full coding scheme and each pilot coded eight transcripts. Finally, after review and discussion of the coded transcripts, the scheme was refined and the entire dataset coded.

Results

The following results are organised around three interrelated themes: 1) maternal health is shaped at the interface of social and ecological systems; 2) maternal health is a holistic concept and more than an issue of reproductive health; and 3) the intergenerational reproduction of embodied inequalities. Direct quotations from in-depth interview (IDI) and focus group transcripts (FGs) are used to punctuate these themes and contextualise participants' responses. Each quotation is identified with a pseudonym, data collection technique (i.e., FG or IDI) and the year of data collection.

Maternal health as a social and ecological issue

While women indicated the MEP was an important factor for maternal health, they commonly discussed how their social and physical environments were important drivers in their maternal and overall health. Certainly, the MEP provides the psychological assurance of having health coverage; many women indicated how their health, and that of their children, was intimately connected to and dependent upon their ability to generate sufficient financial resources. Throughout discussions it became clear that women's social realities and livelihood opportunities informed their perceptions and experiences of maternal health and the health of their children:

> Women may be sick but you have to go to the farm and work, so unless it's so severe, you might not think it's a priority.
>
> (Esther, IDI, 2011)

> I go to look for firewood to go and sell. Sometimes you go into the forest and spend five, six, seven hours to look for firewood and you come and sell it for two cedis but it's from that that I have to feed the family, take care of myself and pay for this child here. If we can't eat, they're hungry and get sick.
>
> (Rashita, FG, 2013)

Equally important was the observation that the land, the physical environment, was a critical component of health. Women often discussed how the changing physical climate and environmental conditions of the UWR were straining their ability to make any form of livelihood and provide adequate food for themselves and their children. Many women suggested that increased drought and poor soils hindered sufficient seasonal crops, resulting in limited food, both to eat and sell, reinforcing poor health outcomes for both women and children:

> If we can't grow food, we don't eat, then the child aren't healthy... A lot of the children we give birth to are malnourished.
>
> (Annolata, IDI, 2011)

> We [women] work the land. That's how most of us make any kind of livelihood. But now the soils are losing nutrients, the rains are sporadic and the crops don't grow. It's hard because that's what we eat and what we sell, so you're stuck. It makes life hard. It's okay if I go without food but if the kids do, it's tough.
>
> (Juba, FG, 2013)

> [Looking over time] the living standards these days are better, but today most women can't even afford to take care of their kids. But those days there was abundance of food and everything even before you get pregnant. Your husband would fill the whole place with food and everything. But today, despite the fact that healthcare [insurance] is improving, women are suffering to take care of their children [provide food].
>
> (Thelma, IDI, 2013)

Similarly, to understand maternal healthcare in the UWR is to realise women's health needs occur in deprived contexts with limited economic opportunities, primarily based on subsistence farming, fashioned to meet familial obligations and household responsibilities. This means resources to (potentially) pay for maternal health needs (e.g., transport, medicine, personal items) must first be divided among household members and child necessities. Competing obligations and "compulsory" workloads to provide economically, feed their children and tend to other household tasks often took priority over maternal care:

> I have to leave all of my children at home when I go to the hospital. If they don't have food, it's hard... because I would be working instead of going to the facility.
>
> (Angelina, FGD, 2013)

> You're forced to weigh your needs. Do you want healthcare or would you rather make money and look after the kids? Because if you go, you leave them by themselves. You can't do that because they're small. But even if

you need to [get maternal healthcare], it's going to cost to get there and that money could be used for other things at home.

(Mawusu, IDI, 2013)

Even if you have some small money, why would you use that to get care [maternal healthcare]? I know it's supposedly free but you have to find and pay for transport, then you have to pay for the things at the facility like rags, gauze, Dettol [antiseptic], soap and nappies. That money could've been used to get food, and provide for the children because you know the husbands won't do it. It's tough to make that decision.

(Margaret, FG, 2011)

Though the MEP was assumed to resolve maternal health challenges by providing health coverage, a significant disconnect emerged with how participants described factors contributing to their maternal health. Women consistently stated that maternal health was not just about having insurance, but related to the broader socioeconomic challenges of the region. That is, high levels of poverty and limited socioeconomic opportunity:

Poverty is a sickness. The high levels of poverty, which is still worse among women, isn't accounted for with the health policy. The free maternal health exemption, it's only a small piece for health.

(Kessie, FG, 2013)

They [government] thought that giving us "free healthcare" would make things better but it doesn't. We're still living in the same conditions with very little. The husbands don't care about us. Even if you can get healthcare with the policy we go back to the same conditions we were in and we remain sick.

(Esi, IDI, 2013)

Political agendas and contradictions of maternal health

While the services provided under free health insurance were beneficial to women and their newborn children (i.e., antenatal care, skilled delivery, postnatal care), women commonly discussed their problematic and contradictory implementation. Throughout discussions, women revealed the government had previously provided pregnant women with food supplements during pregnancy to assist in maternal and foetal health, but the programme was discontinued with the implementation of the MEP. In a setting with high levels of food insecurity and anaemia, it was perceived by participants as counterintuitive to fund programmes for MNCH and simultaneously discontinue a food programme for pregnant women:

The health providers used to give us food supplements [iron] when we went to the antenatal and postnatal [appointments], but now it's been stopped. They wanted to encourage women to come to the clinic for delivery but that same year they stopped supplements.

(Margaret, IDI, 2013)

I don't understand why they stopped giving us the [iron food] supplements, that's what we need because a lot of us are anaemic. Because of that, we can't properly feed our kids. What kind of milk is that? We don't produce anything of nutrition. The [kids] now don't grow like they used to.

(Abla, FG, 2013)

Foodstuff is hard to come by. We're lucky if we can get salt and sugar for cooking let alone any meat. It's just broth and some small vegetables if we can. So I think that makes us weak and worse when we have kids. We bleed a lot and it takes so long to stop. Since we don't eat properly when we're pregnant, the kids don't grow so well. When they're born, they're underweight.

(Medius, FG, 2013)

Still, when women went to the health facility for antenatal check-ups, they were told to sit through lectures from local health providers informing them of the nutritious food they should consume during pregnancy. However, few women had access or means to purchase such foodstuff:

These things in the guidelines, it's very hard for us to come by. Most of us give birth and have not eaten any of these things. The government should include the food supplements for us again … supplements should be part of the exemption … it would help to reduce some of the problems, diseases and things.

(Fanny, FG, 2013)

Many women made links between their poor nutrition, high rates of anaemia and challenging experiences giving birth. Discussing issues related to birth, numerous women indicated losing substantial amounts of blood, often becoming very weak and jeopardising the health of the women and (un/newly) born child. In these cases, women suggested high rates of anaemia from inadequate nutrition worsened issues of high blood loss during birth:

You know, women giving birth lose a lot of blood and you may need a blood transfusion but that's not covered [in the MEP], and the money's not easy to come by…

(Josephine, FGD, 2013)

Because we don't have enough iron, a lot of us [women] are anaemic. Then when we give birth, we lose a lot of blood and it takes so long to stop. Some even need transfusions but there are shortages of blood and it's costly. It's a risk when you give birth, but there's nothing you can do.

(Prossie, IDI, 2011)

The nurses say we should be eating all of this food, like meat and fruits and vegetables, but we can't get any of or even afford it if it was here. We can't grow a lot of food here because the soils are poor and the quality of the land has deteriorated.

(Adina, IDI, 2013)

Intergenerational reproduction of embodied inequality

Throughout discussions, it became clear that very few women received any economic support from their husbands for their maternal healthcare or healthcare for their children. MNCH was often deemed a woman's "problem":

Women are marginalised in the first place. Among the poor we are the poorest ... Sometimes looking at the way women are marginalised, even if you say you're married the husbands won't even take care of you, so you have to take care of yourself and the children.

(Addae, IDI, 2013)

We women always end up paying to enrol the children and ourselves. You know, women are usually susceptible to sickness and other diseases that are just only for women, which are always common, especially here. [And then] if you have a child and the child is sick, it is actually you, the woman, who takes care of the child. They do everything with you so that is the most reason why women need good health.

(Mawusu, FGD, 2011)

The introduction of the MEP thus intended to solve the problem by providing "free" maternal healthcare for a year. Yet, the assumption that issues of individual reproduction could be resolved with a biomedical fix ignored the underlying causes of poor maternal health outcomes. Indeed, many women pointed to broader underlying causes of poor maternal and child health rather than medical-based solutions:

In these parts of Ghana, the poverty level is very high, people find it hard to even feed themselves, but childbirth is a natural right for women, you can't say because of poverty you won't give birth, if you're poor or you're rich it's every woman's right to give birth. So, I think even the exemption policy isn't enough because there's a lot involved in childbirth. If there was something else, especially at the community level, at least food supplements and soap, those things would even be more helpful and give a truer meaning to maternal health than just the free health exemption.

(Ekua, FG, 2013)

When discussing the health of women and children, many alluded to the concept of a cycle, between themselves, the environment and the health outcomes of their children. Links were often made regarding changing livelihood opportunities including droughts, poor crops, limited economic activity, their poor nutrition and high rates of anemia, and the health of their children:

A lot of the children we give birth to are malnourished. I know, we don't eat well, so they get sick. How are they to grow?

(Maloo, FG, 2011)

You see, the woman's health affects the child. We don't have enough food and the crops are bad. There's no help and a lot of the children we give birth to are malnourished.

(Kakua, IDI, 2013)

Together, women's marginalised status and meagre livelihood opportunities positioned women within a precarious situation of uncertain health outcomes for both women and children, creating an intergenerational cycle of poor maternal health and infant-child health outcomes:

We work hard, us women. We work the farms and try to make some kind of living. The harvests used to be so good, but now it's hard to grow enough food. The soils are changing and have lost a lot of nutrients. We try our best but it's never enough. Then our husbands don't help us. We have to manage the household, the kids, and find food but it's never enough .... We work long hours and are often weak. We're lucky if we have some salt for soup. Meat is a luxury. We hardly ever get that so we're not really strong, especially during pregnancy. That consumes much energy so we end up producing babies that aren't well. It's a never-ending cycle.

(IDI, Adina, 2013)

It would be good if the health insurance scheme took care of us women and insured our children. If you're going to struggle to get enough money, they should reduce the cost. Even if they can't extend it for the women, it should be extended for the children, so they can have proper healthcare.

(Abla, IDI, 2013)

To summarise, a critical point from these narratives is that without attending to the multiplicity of environmental, social, historical and political-economic contexts of the UWR, it will be difficult to understand why women's maternal health remains such a large concern.

Discussion and conclusion

While the MEP is a significant factor promoting MNCH, we must be cognisant that maternal health is more than simply having insurance or access to skilled delivery in a health facility. The theoretical vantage point of feminist political ecology of health allows us to ask different questions and uncover "subaltern health narratives" (King 2010, 14). Challenging the assumption that a maternal health policy will automatically translate into improved maternal health outcomes motivated the two central questions of the chapter: (1) What are the factors other than insurance contributing to maternal health? (2) What are the gendered and embodied experiences of the environment that contribute to maternal and child health? By scaling down our analysis to examine how women's daily experiences are shaped through the historical political economy of the UWR, we elucidate how gender roles and power structures shape human environmental dimensions of maternal health (Sultana 2012).

Foremost, maternal health depends *not* just on having health insurance (i.e., access to a health facility with a skilled attendant), but on a gamut of social and environmental resources necessary to meet maternal health needs. Our finding suggests that in the UWR, where women's maternal health needs compete with a constellation of other factors (farming, child-rearing, household responsibilities, et cetera), issues of health become a second thought. For instance, women's abilities to define access to and control over resources (finances, food production), were interrelated to gendered and cultural factors embedded in the UWR. The fact that women often neglected maternal health concerns in lieu of economic sustenance, or the need to acquire food for their children while going hungry themselves, underscores how opportunities for maternal health extend far beyond access to medical expertise. These are the social and political elements embedded in everyday life within a patriarchy, all of which impinge on MNCH. These findings resonate with many studies in feminist political ecology, elucidating how gender politics shape access to resources (Rocheleau et al. 1996).

Beyond the political arena of the household, these results speak to the ways the physical environment intersects with maternal health. Women made links between the poor soil quality, sporadic rainfall and variabilities in climate with their nutrition, physical and psychosocial health, highlighting the ways the biophysical environment meets with the human body to produce expressions of (ill) health. Women's access to natural resources, land allocation and food sources all intersect with biophysical processes to shape the trajectories of health and wellbeing. These findings buttress other work from the UWR showing food insecurity as detrimental to physical and psychosocial health outcomes (Atuoye and Luginaah 2017).

Similarly, these results speak to the ways women's gendered position and often marginalised status become embodied and reproduced in infant-child health outcomes. In a way, women's limited economic resources, lack of autonomy and constrained ability to acquire sufficient caloric and nutritional demands were physically expressed in the health outcomes of infants and children. This, in turn, perpetuates a cycle of poor health. These dynamics emphasise how health is a coupled social and ecological process (King and Crews 2013).

More broadly, situating women and their communities within broader political and economic environments demonstrates the ways MNCH agendas traverse material environments. Our results provide a snapshot into the ways health policies are implemented in practice. In an era of scarce healthcare resources, the government decided to stop providing food and iron supplements the same years the MEP was implemented. Execution of the MDG goals are certainly not always perfect, and indeed fragmented and siloed approaches are often taken towards maternal health (Graham et al. 2016; Horton 2016). The assumption that "health" would be guaranteed through insurance coverage, without recognising the local practices and knowledge systems embedded in the UWR, led to detrimental health outcomes for women and children. This finding draws attention to the ways in which political agendas inform women's lived realities and bodily expressions of health as they struggle to sustain ecologically viable livelihoods (Rocheleau et al. 1996).

Similarly, while gender equity is cited as a goal of the MEP, policies which address the promotion of gender equitable healthcare must align with the distribution of health-related concerns (Langer et al. 2015). As our findings reveal, the design of the MEP only considers women's maternal health needs for a one-year period, suggesting women's health is insignificant beyond child birth. Hence, defining women's health primarily in terms of "reproduction" (i.e., fertility, pregnancy, childbirth) neglects the full spectrum of women's health. With little consideration of women's local realities, these political decisions perpetuate deleterious health outcomes. In particular, the perceived rivalry between healthcare and purchasing food highlights how MNCH is a product of spatial economies that have been produced over time by the convergence of political, economic, social, gendered, cultural and ecological systems that intersect and create embodied expressions of maternal and child health and wellbeing (see King 2017). Overall, the findings illuminate how historical, social and ecological processes create new health challenges—the consequences of which are lived and experienced by gendered bodies. This speaks powerfully to the literature on PEH, which shows how environmental, social and historical contexts converge in complicated ways to produce negative health outcomes (King and Crews 2013; King 2015; Neely 2015).

Conceptually, the case study extends the literature in the areas of gender and embodiment in PEH, showing not just why bodies matter (Hayes-Conroy and Hayes-Conroy 2015), but why PEH scholars must be attentive to politics inscribed in various social arenas, including households, neighbourhoods and villages. Theoretically, "the realm of bodily experience can and should be included in all political ecology theorizing" (ibid., 659). We echo this call and suggest more work not just on bodily experience but also richer case studies on gendered bodily experiences. Indeed, this will immeasurably expand the multi-scalar analysis of PEH and incorporate the multifaceted axes of difference (Elmhirst 2015).

To conclude, we return to our overarching question: are maternal and child health programs in the era of the SDGs tipping into irrelevance? The evidence we have presented in this chapter suggests the answer is at best uncertain. However, we do not think this means giving up the implementation of MNCH policies and programs in low and middle income countries. We would like to emphasise the progress Ghana and other low and middle income countries have made in improving measured outcomes, despite the often convoluted implementation of these policies (Storeng and Béhague 2014; Graham et al. 2016). However, many challenges still lie ahead for MNCH policies and programs in the SDG era. Broader frameworks are needed to encompass concepts of women's health founded on a life course approach, considering factors—social, political, environmental, cultural (i.e., the socio-ecological determinants)—that interact with biology and gender beyond their reproductive years. As major economic, environmental, demographic and epidemiological transitions unfold worldwide, the implications for new health policies to address the health of populations, particularly women and children, must recognise the ways human health is intrinsically connected to biophysical and sociopolitical environments (Gatrell and Elliot 2014).

This means asking new questions beyond the access, enrolment or utilisation of healthcare. For instance, asking why women encounter complicated pregnancies or considering the reasons behind high maternal morbidity and mortality. Doing so would expose important social, biological and environmental systems that perpetuate and compound MNCH challenges; which, in most cases, begin long before conception and are conditioned by access to nutrition, secure livelihoods and economic stability. So, without interrogating the underlying environmental, political, economic, social or cultural circumstances influencing a woman, for example, to give birth on the side of the road, not receive proper nutrition, or broadly put stress on her body, then we are not going to make much of a difference to women's maternal health on the whole. Providing health insurance coverage alone does not mean a safe and healthy delivery. Understanding maternal health in its totality is the challenge going forward, both globally and locally.

Acknowledgements

Our deepest gratitude goes to the women of the Upper West Region who sacrificed their time to participate in this research. To Dr. Isaac Luginaah, thank you for all that you have done for us. This work would not have possible without you.

References

Agarwal, Bina. 1997. "'Bargaining' and Gender Relations: Within and Beyond the Household." *Feminist Economics 3(1)*: 1–51.

Atuoye, Kilian Nasung, and Isaac Luginaah. 2017. "Food as a Social Determinant of Mental Health among Household Heads in the Upper West Region of Ghana." *Social Science & Medicine 180*: 170–180.

Dixon, Jenna. 2014. "Determinants of Health Insurance Enrolment in Ghana's Upper West Region." PhD diss., University of Western Ontario.

Dzakpasu, Susie, Seyi Soremekun, Alexander Manu, Guus ten Asbroek, Charlotte Tawiah, Lisa Hurt, Justin Fenty *et al.* 2012. "Impact of Free Delivery Care on Health Facility Delivery and Insurance Coverage in Ghana's Brong Ahafo Region." *PloS One 7(11)*: e49430.

Elmhirst, Rebecca. 2015. "Feminist Political Ecology." In *The Routledge Handbook of Political Ecology*, edited by T. Perreault, G. Bridge, & J. McCarthy, 519–530. New York: Routledge.

Ewusie, Joycelyne E., Clement Ahiadeke, Joseph Beyene, and Jemila S. Hamid. 2014. "Prevalence of Anemia among Under-5 Children in the Ghanaian Population: Estimates from the Ghana Demographic and Health Survey." *BMC Public Health 14(1)*: 626.

Fehling, Maya, Brett D. Nelson, and Sridhar Venkatapuram. 2013. "Limitations of the Millennium Development Goals: A Literature Review." *Global Public Health 8 (10)*: 1109–1122.

Ganle, John Kuumuori, Isaac Dery, Abubakar A. Manu, and Bernard Obeng. 2016. "'If I Go with Him, I Can't Talk with Other Women': Understanding Women's Resistance to, and Acceptance of, Men's Involvement in Maternal and Child Healthcare in Northern Ghana." *Social Science & Medicine 166*: 195–204.

Gatrell, Anthony. C., & Elliott, Susan. J. 2014. *Geographies of Health: An Introduction*. Malden, MA: John Wiley & Sons.

Glover-Amengor, Mary, Isaac Agbemafle, Lynda Larmkie Hagan, Frank Peget Mboom, Gladys Gamor, Asamoah Larbi, and Irmgard Hoeschle-Zeledon. 2016. "Nutritional Status of Children 0–59 Months in Selected Intervention Communities in Northern Ghana from the Africa RISING Project in 2012." *Archives of Public Health 74(1)*: 12.

GNHIA (Ghana National Health Insurance Authority). 2013. *2013 Annual Report*. Accra, Ghana: Ghana National Health Insurance Authority.

Graham, Wendy, Susannah Woodd, Peter Byass, Veronique Filippi, Giorgia Gon, Sandra Virgo, Doris Chou *et al.* 2016. "Diversity and Divergence: The Dynamic Burden of Poor Maternal Health." *The Lancet 388(10056)*: 2164–2175.

GSS (Ghana Statistical Service). 2012. *Ghana Multiple Indicator Cluster Survey with an Enhanced Malaria Module and Biomarker. Final Report*. Accra, Ghana: Ghana Statistical Service.

GSS (Ghana Statistical Service). 2013. *2010 Population and Housing Census Report: Non-Monetary Poverty in Ghana*. Accra, Ghana: Ghana Statistical Service.

Guthman, Julie. 2012. "Opening up the Black Box of the Body in Geographical Obesity Research: Toward a Critical Political Ecology of Fat." *Annals of the Association of American Geographers 102(5)*: 951–957.

Guthman, Julie, and Becky Mansfield. 2013. "The Implications of Environmental Epigenetics: A New Direction for Geographic Inquiry on Health, Space, and Nature-Society Relations." *Progress in Human Geography 37(4)*: 486–504.

Guthman, Julie, and Becky Mansfield. 2015. "Nature, Difference and the Body." In *The Routledge Handbook of Political Ecology* edited by T. Perreault, G. Bridge, and J. McCarthy, 558–570. New York: Routledge.

Hayes-Conroy, Jessica, and Allison Hayes-Conroy. 2015. "A Political Ecology of the Body: A Visceral Approach." In *The International Handbook of Political Ecology*, edited by Raymond L. Bryant, 659–672. Cheltenham, UK and Northampton, MA: Edward Elgar.

Hessey-Biber, Sharlene. 2003. *Approaches to Qualitative Research: A Reader on Theory and Practice.* Oxford: Oxford University Press.

Horton, Richard. 2016. "Offline: Global Health—Tipping into Irrelevance." *The Lancet 388(10052)*: 1362.

Kabeer, Naila. 2015. "Tracking the Gender Politics of the Millennium Development Goals: Struggles for Interpretive Power in the International Development Agenda." *Third World Quarterly 36(2)*: 377–395.

Kassebaum, Nicholas J., Ryan M. Barber, Zulfiqar A. Bhutta, Lalit Dandona, Peter W. Gething, Simon I. Hay, Yohannes Kinfu *et al.* 2016. "Global, Regional, and National Levels of Maternal Mortality, 1990–2015: A Systematic Analysis for the Global Burden of Disease Study 2015." *The Lancet 388(10053)*: 1775–1812.

Kassebaum, Nicholas J., Rafael Lozano, Stephen S. Lim, and Christopher J. Murray. 2017. "Setting Maternal Mortality Targets for the SDGs–Authors' Reply." *The Lancet 389(10070)*: 697–698.

King, Brian, and Kelley A. Crews, eds. 2013. *Ecologies and Politics of Health.* Abingdon and New York: Routledge.

King, Brian. 2010. "Political Ecologies of Health." *Progress in Human Geography 34 (1)*: 38–55.

King, Brian. 2015. "Political Ecologies of Disease and Health." In *The Routledge Handbook of Political Ecology*, edited by T. Perreault, Gavin Bridge, and James McCarthy, 343–353. New York: Routledge.

King, Brian. 2017. *States of Disease: Political Environments and Human Health.* Oakland, CA: University of California Press.

Jackson, Paul, and Abigail H. Neely. 2015. "Triangulating Health: Toward a Practice of a Political Ecology of Health." *Progress in Human Geography 39(1)*: 47–64.

Kidd, Pamela S., and Mark B. Parshall. 2000. "Getting the Focus and the Group: Enhancing Analytical Rigor in Focus Group Research." *Qualitative Health Research 10(3)*: 293–308.

Langer, Ana, Afaf Meleis, Felicia M. Knaul, Rifat Atun, Meltem Aran, Héctor Arreola-Ornelas, Zulfiqar A. Bhutta *et al.* 2015. "Women and Health: The Key for Sustainable Development." *The Lancet 386(9999)*: 1165–1210.

Miles, Matthew B., A. Michael Huberman, and Johnny Saldana. 2014. *Qualitative Data Analysis: A Method Sourcebook.* Thousand Oaks, CA Sage Publications.

McDougall, Lori. 2016. "Discourse, Ideas and Power in Global Health Policy Networks: Political Attention for Maternal and Child Health in the Millennium Development Goal Era." *Globalization and Health 12(1)*: 21.

Neely, Abigail H. 2015. "Internal Ecologies and the Limits of Local Biologies: A Political Ecology of Tuberculosis in the Time of AIDS." *Annals of the Association of American Geographers 105(4)*: 791–805.

Nyantakyi-Frimpong, Hanson, and Rachel Bezner Kerr. 2017. "Land Grabbing, Social Differentiation, Intensified Migration and Food Security in Northern Ghana." *The Journal of Peasant Studies 44(2)*: 421–444.

PMNCH (Partnership for Maternal, Neonatal and Child Health). 2009. *Partnership Vision: Invest, Deliver, Advance.* Available at: http://www.who.int/pmnch/en

Richmond, Chantelle, Susan J. Elliott, Ralph Matthews, and Brian Elliott. 2005. "The Political Ecology of Health: Perceptions of Environment, Economy, Health and Well-being among 'Namgis First Nation." *Health & Place 11(4)*: 349–365.

Rishworth, Andrea C. 2014. "Women's Navigation of Maternal Health Services in Ghana's Upper West Region in the Context of the National Health Insurance Scheme." PhD diss., University of Western Ontario.

Robbins, Paul. 2011. *Political Ecology: A Critical Introduction.* Malden, MA and Oxford: John Wiley & Sons.

Rocheleau, Dianne, Barbara Thomas-Slayter, and Esther Wangari, eds. 1996. *Feminist Political Ecology, Global Issues, and Local Experiences.* London: Routledge.

Rocheleau, Dianne E. 2008. "Political ecology in the key of policy: From chains of explanation to webs of relation." *Geoforum 39(2)*: 716–727.

Songsore, Jacob. 2003. *Regional Development in Ghana: The Theory and the Reality.* Accra: Woeli Pub. Services.

Starrs, Ann M. 2014. "Survival Convergence: Bringing Maternal and Newborn Health Together for 2015 and Beyond." *The Lancet 384(9939)*: 211–213.

Storeng, Katerini T., and Dominique P. Béhague. 2014. "'Playing the Numbers Game': Evidence-based Advocacy and the Technocratic Narrowing of the Safe Motherhood Initiative." *Medical Anthropology Quarterly 28(2)*: 260–279.

Sultana, Farhana. 2012. "Producing Contaminated Citizens: Toward a Nature–Society Geography of Health and Well-being." *Annals of the Association of American Geographers 102(5)*: 1165–1172.

Takyi, B.K. & Oheneba-Sakyi, Y., eds. 2006. *African Families at the Turn of the 21st century.* Westport, CT: Praeger Publishers.

Tolhurst, Rachel, Yaa Peprah Amekudzi, Frank K. Nyonator, S. Bertel Squire, and Sally Theobald. 2008. "'He Will Ask Why the Child Gets Sick so Often': The Gendered Dynamics of Intra-household Bargaining over Healthcare for Children with Fever in the Volta Region of Ghana." *Social Science & Medicine 66(5)*: 1106–1117.

Tsikata, Dzodzi, and Joseph Awetori Yaro. 2014. "When a Good Business Model Is Not Enough: Land Transactions and Gendered Livelihood Prospects in Rural Ghana." *Feminist Economics 20(1)*: 202–226.

UNSEA (United Nations Social and Economic Affairs). 2015. *Global Sustainable Development Report, 2015 Edition.* New York: United Nations.

WHO (World Health Organization). 2012. *Trends in Maternal Mortality 1990–2010.* Geneva: WHO.

Conclusion

Reproductive bodies, places and politics: Future directions

Helen Hazen, Marcia R. England and Maria Fannin

In this volume, we focus on the sites, spaces and subjects of reproduction in order to expand our understanding of the political, cultural and material dimensions of fertility, pregnancy and birth. Bringing together new empirical, methodological and theoretical approaches, *Reproductive Geographies* demonstrates the spatial aspects of reproduction. This geographic lens suggests that bodies, places and politics are key frames of analysis that can reveal important tensions related to the processes of fertility, pregnancy and birth. One of the goals of the text is to become more attentive to how understandings of these frames can improve or compromise the lived experiences of reproduction.

Part I focused on the *body* and how experiences and understandings of reproduction, including social, cultural and political processes, become embodied. The body is a key scale at which reproductive experiences and technologies are played out. Feminist analyses of the regulation of the reproductive body, as well as cultural imaginaries and material practices surrounding women's bodies, reveal the significance of gendered presumptions about experiences of reproduction. In this light, Part I suggests how a geographical analysis of the body and embodiment can further research on reproduction by examining new and varied sites of reproductive practice including biobanks, clinics and urban infrastructures.

Our geographic focus drew attention to *place* in Part II. Places are constantly changing, as are people's navigation of them, and this is particularly significant when new technologies and evolving cultural practices create constantly changing landscapes of reproduction. New sites of birth, associated with changing ideas of safety and risk, provide a traditional example of the significance of place to reproduction. However, in this volume, we push notions of place further as technologies such as artificial insemination open up novel reproductive sites of analysis as more and more babies can be envisaged as being "created" in the clinic, in a lab, or *in vitro*. How one understands place is dynamic and one's changing reproductive status can greatly influence that. The chapters in this section argued that attention to place is essential in understanding the changing landscapes of reproduction.

Part III focused on the dialectical relationship between reproduction and *politics*, including political ideologies. Reproduction is so integrally inter-twined with population size and characteristics as well as resource issues that it has become a central political issue. Reproduction is further politicised as nation-building projects become tied to particular visions of a country's reproductive history and future. This section therefore explores how move-ments and conceptions of citizens and non-citizens have become important in understanding reproductive geographies. Although political geographers have studied these issues for many years, we argue that additional value can be gained by viewing these issues through a reproductive as well as political lens.

In summary, *Reproductive Geographies* brings together multiple perspec-tives, theoretical frameworks and empirical sites to show how bodies, places and politics are important dimensions of (pre)conception, pregnancy, surro-gacy, birth and the perimenopause. We include contributions focused on phenomenologically inspired accounts of women's lived experience of preg-nancy and birth, the biopolitics of birth and citizenship, the material histories of reproductive tissues as "scientific objects" and engagements with public health and development policy. We bring together work that directly addres-ses global perspectives on reproductive politics as well as state-focused approaches to the politicisation of pregnancy and birth. The book aims to inspire further work on how geography is central to reproduction, and repro-duction an important dimension of geographers' concerns, from geopolitics to health to labour. By adding a spatial lens to reproduction through discussions of scale, boundaries and margins, and place, this volume brings a much needed geographical perspective to bear on experiences and ideologies of reproduction.

Gaps and silences

With a view to furthering our understanding of the politics, places and bodily spaces of reproduction, we draw attention to some notable gaps and silences that remain in the literature.

First, considerably more attention has been paid to issues of fertility than experiences of *in*fertility, with pregnancy loss and termination being particu-larly understudied. Pregnancy loss, including miscarriage and termination, is not often researched by geographers, and yet would benefit from exploration as an extremely significant embodied experience. Loss can bring pain, silence and hiding, and lack of research in this area furthers the stigma associated with both miscarriage and termination. Although clearly a challenging area of enquiry, it is imperative that more attention be paid to pregnancy loss in acknowledgement of the large proportion of women who experience a termi-nation or miscarriage at some point in their reproductive lives.

More broadly, issues of infertility raise important questions of social justice as those who have already conceived frequently receive greater protection by law and support from wider society than do those who are unable to conceive "naturally". Women's movements have made considerable efforts to try to protect the pregnant person's right to choose whether or not to continue a pregnancy, and yet this right is out of reach for those who have difficulties with conception in the first place. Should people have the right to explore all avenues to try to conceive, even as new technologies change our understanding of who can, and who cannot, be biological parents? Apart from their obvious ethical dimension, such questions also have significant financial implications, and so we argue that we must extend discussion of reproductive justice in light of these new technologies. Such discussions might include: subsidising donor banks, better laws protecting the rights of families using known donor sperm, improving the experience of single women and same-sex couples with reproductive medicine, and funding more research into emerging technologies. Raising these issues is not to detract from the significance of more traditional reproductive debates, such as access to contraception and pregnancy termination, and may even lead to further insights into some of these longstanding and somewhat intractable debates.

Second, queer geographies of reproduction also remain underexplored in the geographical literature. Historical and current narratives often marginalise, or even ignore, LGBTQ+ people in the fight for reproductive equality. As a first step towards addressing such inequalities we need more diverse accounts of "reproduction" in geography than those of the white, heterosexual middle-class woman or couple. The presumption that geographical work on pregnancy and birth focuses primarily on these privileged subjects is supported by the paucity of work exploring diverse geographies of reproductive bodies, places and politics. The reproductive experiences of partners also remain understudied, with most work referring only in passing to the experiences of partners—whether men or women. LGBTQ+ reproductive politics, including adoption laws, access to reproductive technologies, trans pregnancy, and comprehensive sex education, are critical issues in these current and future landscapes of reproduction.

Third, we highlight that much of the work currently available on reproductive geographies focuses on the Global North, with the experiences of women and men of the Global South relatively neglected. Where these contexts are explored, researchers have tended to focus on Southern women's reproductive roles in relation to more affluent countries—as sources of surrogate mothers, for instance—or to view reproduction from a population-level perspective centred on birth rates or other measures of fertility, pregnancy and birth. We have yet to hear many of the day-to-day reproductive experiences of women from the Global South.

Finally, we argue that discussions of population and resources must be extended in a new era of complex political dynamics and potential resource crisis. Climate change, water shortages and landscape conversion raise questions about the relationship between a growing human population and the environment that supports it. Traditionally, scholars have used an ecological lens to explore such issues, and yet the very complexity of these "wicked problems" suggests the need for multiple perspectives to find more just and sustainable ways of living. Faced with an uncertain planetary future, the control of population, with its roots in eugenic thinking, is re-emerging as a matter of public concern. Geographers are well-placed to interrogate these and other pressing problems about the future of life on the planet. Reproductive geographies, with their focus on the entanglements of bodies, politics and places, must surely contribute to these debates.

Index

abnormality of embryos 9, 17–18, 23, 24, 32–40, 42–43, 44n7
abortion 3, 5, 43; in Brazil 2; in Ireland 3; in Japan 17–19, 20–23, 26, 27n3; in Serbia 192–193, 222; in the U.S. 3
"accidental citizenship" 162, 166–168
agency 4, 112, 114–115, 144
alien citizenship 162, 166
Almeling, Rene 52
American College of Obstetricians and Gynecologists (ACOG) 126, 139, 143
American Medical Association (AMA) 143
anatomical collections 9, 17–27
"anchor babies" 166–167, 176
Appalachia 146–147
artificial insemination 8, 48–50, 52–53; and the body 54–56, 59, 63; costs of 54, 59–62; and mobility 51–52, 61–62, 63; process of 52–53; sites of 1–2, 9, 48, 53, 59–61, 63; timing of 48, 52, 53
assisted reproductive technologies (ART) *see* reproductive technologies
autonomy 3, 216; and birth 114, 123, 143, 144, 148, 150–151, 155; political 177, 188, 192–194

"backward uncitizening" 168, 178
Berg, Barbara 3, 4
Bigo, Didier 162, 163, 165, 166
biomedicine 148, 202, 213; and birth 123, **129**, 144, 147–148
biopolitics 2, 7, 11, 27, 38–39, 222; in China and Hong Kong 161–162, 163–168, 172, 177–178, 180n6; in Serbia 184–190, 192, 194–197
biotechnology 31–33, 37, 42
biovalue 33, 40, 42, 43

birth 1, 4, 5, 11, 56, 122, 134, 221–222; biomedical model of birth 123, 144; in China/Hong Kong 161–162, 169–178, **171, 173**; in Ghana 202, 207, 212–214, 122–123, 125–126, 131, 143, 147–148, 151, 153; interventions 125, 126, 129, 144, 147; medicalisation of 6, 124; and mobility 150–151; in Serbia 184–186, 197; and surrogacy 109–112, 115, 117–118; out-of-hospital birth 10, 123, 124, 129–130, **129**; *see also* homebirth; hospital birth
birth centre 10, 122–124, 126, 130–139, 142
birth control 3, 179; in Japan 22, 23
Birth of the Clinic (Foucault) 36, 49
birthplace 1, 145, 146, 155
birthright citizenship 162, 165, 166–169, 171, 179, 180n4, 180n6
body 1, 2, 3, 6–11, 49, 93, 155, 221; fluids 91, 93–94, 150, 179; fragmentation of 3, 17–19, 26, 31-32, 34, 49–50, 52, 59, 62–63, 91; geographies of 1, 2, 19, 32, 49, 142, 145; interior of 7, 17–19, 27, 49; maternal bodies 26–27, 31, 56, 91–93, 95, 99–102, 108, 111, 114–115, 118, 124, 142, 145, 150–151, 153, 185; nationalism and bodies 190–191; and political ecology 204; and politics 163, 165, 167–168, 190; as scientific object 19, 101
breastfeeding 2, 6, 56, 91, 93, 131, 137, 145, 149

Cesarean section (C-section) 129, 144, 146, 147, 207
capitalism 6–7, 111, 115
caring labour 6, 11

For Product Safety Concerns and Information please contact our EU
representative GPSR@taylorandfrancis.com
Taylor & Francis Verlag GmbH, Kaufingerstraße 24, 80331 München, Germany